Fashion Accessories

A Complete Guide to Raw Materials, Construction Methods and Styles

Fashion Accessories

A Complete Guide to Raw Materials, Construction Methods and Styles

Dr. Yamini Jhanji Dhir

CRC Press
Taylor & Francis Group
Boca Raton London New York

CRC Press is an imprint of the
Taylor & Francis Group, an **informa** business

WOODHEAD PUBLISHING INDIA PVT LTD

New Delhi

First published 2024
by CRC Press
4 Park Square, Milton Park, Abingdon, Oxon, OX14 4RN

and by CRC Press
2385 NW Executive Center Drive, Suite 320, Boca Raton FL 33431

CRC Press is an imprint of Informa UK Limited

British Library Cataloguing-in-Publication Data
A catalogue record for this book is available from the British Library

ISBN: 9781032630045 (hbk)
ISBN: 9781032630069 (pbk)
ISBN: 9781032630076 (ebk)

DOI: 10.4324/9781032630076

Typeset in Times New Roman
by Bhumi Graphics, New Delhi

WPI

Contents

Preface

The book aims to provide a holistic and enriched learning experience to readers as far as various designing and technical aspects pertaining to fashion accessories are concerned. Accordingly, the construction methods, materials and styles of varied accessory categories are extensively covered in the book.

The book is intended for undergraduate and postgraduate students, research scholars, academicians and industry professionals associated with textiles, garments, fashion, apparel and design domain. Hence, the course content is designed and tailored pragmatically to suit the requirements of all stakeholders covering wide span coverage of topics beginning from terminologies relevant to fashion world, apparels and fashion accessories, traditional and modern techniques of accessory design and development along with description of accessory brands and styles, smart wearable and sustainable fashion accessories.

1 Fashion and Accessory Terminology

1.1 Fashion

Fashion can be described as a form of self-expression; it is the style or styles that are most popular at a given time. It can be considered a vehicle of conspicuous consumption and upward mobility for fashion conscious consumers to imitate styles of dresses, accessories, hair styles and draping styles from their style icons – celebrities and other famous personalities in limelight. Fashion can be described as a random creation that dies as an innovation is born. Fashion is prevailing or trending - custom or style of attire, accessory, etiquette, socializing, manner or mode. The word fashion is closely associated with customs and trends[1]. Fashion explains relationship between the designed product, its distribution and consumption. The term implies three aspects: style, acceptance and timeliness. Style is any particular characteristic or look in the apparel or accessories. Round neck, V-neck and Band neck are three different styles of one basic upper torso garment, i.e., t-shirt. Likewise, palazzo, pants and culottes are different styles of lower torso garments. Shift dress, A-line, sheath, empire dress are different styles of dresses. A handbag can be designed as per different styles like clutch, tote and envelope bag. Sneakers, ballerinas and thongs are the different footwear styles. Style may come and go in fashion, but a specific style always remains the same, whether it is in fashion or not.Fashionable attire embodies the latest aesthetics, design details, appropriate selection of colors, motifs, prints, trims and notions which capture the mood of an individual and flavor of the season and is desirable at a given moment. Mix and match of entities – upper and lower torso apparels and accessories combined together can be defined as a fashionable look. The concept of buying separates to mix and match instead of buying complete ensembles has augmented the separate purchases of jackets, trousers, shirts, blouses and different accessories to complement the look. The advent of the concept of separates coincided with the advent of the desired casual look.

Figure 1.1 Different dress styles

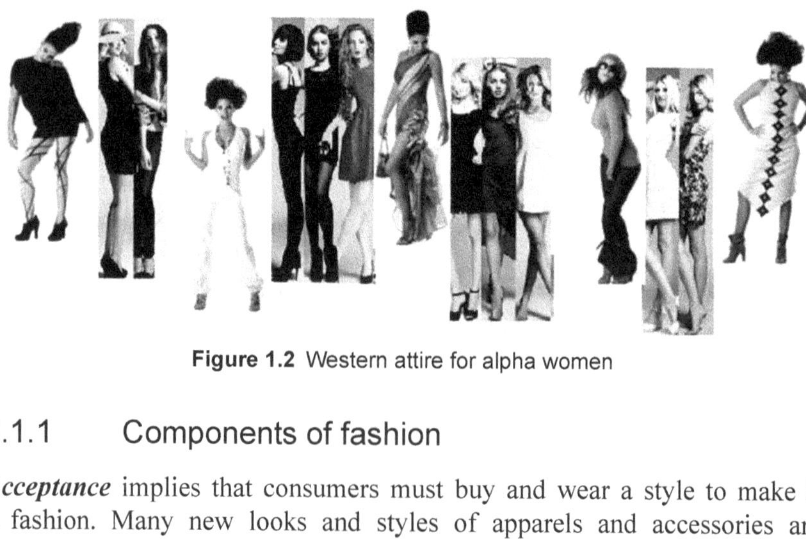

Figure 1.2 Western attire for alpha women

1.1.1 Components of fashion

Acceptance implies that consumers must buy and wear a style to make it a fashion. Many new looks and styles of apparels and accessories are launched and marketed, however, all of them do not readily become staple products owing to reluctance of consumers to accept them. Acceptance of newly introduced trend or style by masses depend on several factors like cost effectiveness, utility, fulfillment of cultural, social and geographical requirements as well as the individual preference of consumer while seeking for anything new in their closets.

Timeliness indicates change. Change is what makes the fashion industry boom and even more exciting and presents ever green business propositions. Tastes and liking of consumers change very abruptly just like changing weather, and the existing apparels and accessories may now be considered as a passe in urge to opt for something unique. It can thus be stated that fashion is a product of change and hence anyone involved with product development or marketing in the fashion industry should have a keen sense of timing, the ability to understand the speed of acceptance and factors affecting the rejection of a particular trend. Fashion involves change, novelty, the context of time, place and most importantly the wearer.

Impetus for fashion is provided by (a) capturing the mood, expression and spirit of individuals from all walks of life in the designer's creations; (b) simultaneous introduction, presentation and display of varied new styles of apparels and accessories; (c) acceptance of newly introduced style by the innovative and fashion-conscious consumer; and (d) updating, introducing and recreating the existing styles analyzing the market trends[2].

Figure 1.3 Silhouettes for women evening gowns

The components of fashion include silhouette, texture, color and details. Silhouette is the outline or shape of a garment. The basic silhouettes preferred for female garments are straight or tubular, bell-shaped, and the bustle or back-fullness. The most desirable silhouette is an hourglass for female wardrobes.

Lines can be straight or curved. Seam lines and lines in patterns give illusion to enhance the personality of the wearer i.e., vertical lines flatter a petite woman more than horizontal lines. Asymmetrical lines, such as chevron-striped fabric in a garment can also be flattering to a short woman, based on their placement. A garment's silhouette plays a pivotal role in determining and enhancing the overall personality and flattering the wearer.

Texture is another element of fashion design. The weave and texture of a fabric has an impact on the way it drapes, which, in turn, affects the way a garment looks when it is worn. Texture relates to the feel of the fabric or the sensation one gets on touching the fabric surface. The texture gets affected by the fiber and the fabricationmethod i.e., knitting, weaving.Fibers, yarns and the method of fabric construction can determine texture. Descriptions of texture includes smooth, rough, dull, shiny, delicate, fine, shaggy and so on. Bulky fabrics have a crucial role in creating illusion of visual appearance and disguising any flaw. Texture undoubtedly plays a role in selection of apparels and accessories by consumers. Rough texture and rugged appearance is preferred while buying a pair of jeans while chiffon with soft feel and shiny appearance that drapes well on body contours is a suitable choice for evening gowns and sarees. The grainy texture and appearance of leather makes it distinct from imitation leather and rexin. Smooth, flat textures give illusion of timid appearance while shiny textured fabric, owing to light reflection, makes an individual look bigger. However, dull textures give the illusion of smaller frame of an individual, as fabrics with dull texture tend to absorb light.

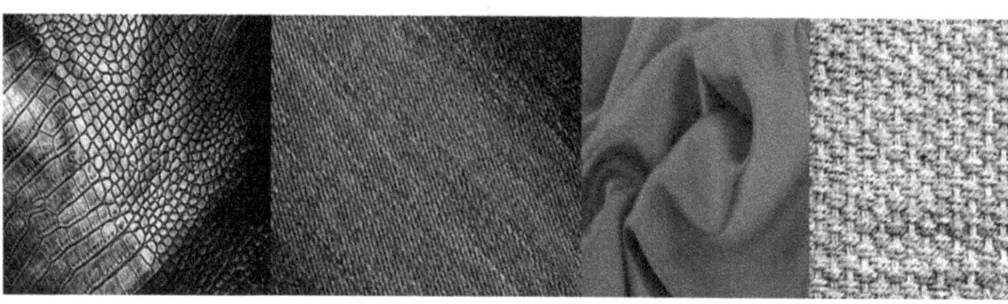

Figure 1.4 Different fabric textures

Color is another important component of fashion, and there are many different color schemes that may work well. Dimensions of color are hue - the quality, value - the lightness or darkness, intensity/chroma - the brightness or dullness, how the surface of a material or a fabric, feels and looks. Neutral colors play a crucial role in apparel and accessory industry as they can be complimented with almost any apparel or accessory, making them good wardrobe-building colors. White, black and grey colors are considered to be neutral colors. Dark colors like blacks and blues usually give illusion of slim and tall appearance, a desirable effect for petite women[3]. Likewise, climate is of paramount importance when it comes to color selection for apparels and accessories. Cool colors – light shades like white and pink are preferred for summers while warm colors which are good absorbers of heat are favorable choice for winter attires. Warm colors are vivid and energetic, and tend to advance in space. Cool colors give an impression of calm, and create a soothing impression. The most common version of color wheel is a wheel of 12 colors, based on the RYB (or artistic) color model. Traditionally, there are a number of color combinations that are considered especially pleasing. These are called color harmonies or color chords and they consist of two or more colors with a fixed relation in the color wheel.

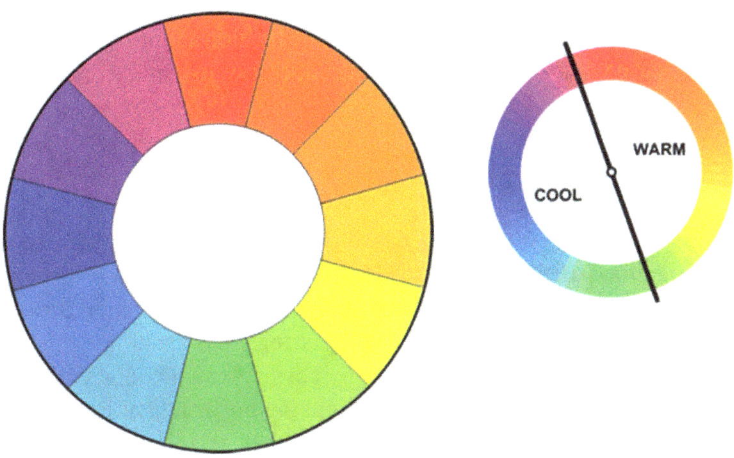

Figure 1.5 Color wheel

1.1.2 Fashion adoption process

The impact of fashion has been influenced by various trends in the society. Different eras have witnessed many popular styles and trends. Designers come up with a gamut of new styles for consumers each season.The acceptance and adoption of designer's creativity and innovation results from individuals making a decision to purchase and wear a new fashion. The fashion adoption is a process throughwhich a new style is adopted by the consumers once the product is commercially introduced in the market. This process involves five basic stages: awareness, interest, evaluation, trial and adoption. The individual becomes aware of the fashion, develops taste and liking for that style and evaluates it in terms of some relative advantages. Relative advantage is the degree that the newly introduced style is seen as better than the previous alternatives, in terms of functionality, cost, social prestige or appealing aesthetics. The consumers often seek for compatibility and complexity in the newly introduced style. Compatibility is the degree to which the style is consistent with the existing norms, rituals, cultural and social beliefs and values of the potential adopters. An innovation is less likely to be adopted if it lies outside the regime of consumer's social beliefs. Likewise, high-level of complexity in selection of color, design details, trimmings and draping styles by designers may not go well with a conservative consumer with traditional outlook. The fashion conscious consumers once satisfied that the style is definitely a better, modified and different version from what is already in their possession, are ready to experiment and go for trails. Positive evaluation of fashion by the individuals leads to the process of trial and adoption. However, observability is also a basic requirement in the fashion adoption process so as to ensure the ease with which the newly introduced style may be communicated, marketed and made available to masses.

1.2 Fashion cycle

The pattern of change in fashion based on consumer acceptancewith time is usually described as the fashion cycle. It is a cycle by which a fashion trend is introduced into the market, the product becomes famous, and then there is a decline in the demand and finally gets rejected. Fashion cyclebasically studies the pattern of fashion adoption by consumers and depicts the rate and time involved in the diffusion process, with the horizontal axis indicating the time and the vertical axis indicating the number of adopters or users. Such graphically portrayed data is useful to calculate the level of acceptance for a fashion. For example, the curve for a fashion that is rapidly adopted but also rapidly declines will represent early growth and quick recession of the style.The graph is also useful to identify the type of consumer in terms of when each adopts a fashion within its life cycle. The consumer who adopts the fashion at the beginning of the curve is fashion leader; at the peak, a mass-market consumer or fashion follower; after the peak, a laggard or isolate.

Fashion cycle is depicted as bell-shaped curve encompassing five stages: introduction of new style, rise in popularity, peak of popularity, decline in popularity and rejection. The willingness, eagerness, disliking and ultimate rejection of newly introduced style dictate the length and timing of fashion cycle. Figure 1.6 shows the different phases of fashion cycle. This rotation is true with different trends and styles of apparels and accessories encompassing a wide range such as color blocking and boyfriend jeans, jumpsuits and shift dress, loafers and sneaker footwear, beanies and cowboy hats.

Some styles may be fads whichdisappear from the market very quickly, while other trends make a comeback and are there for long to stay in the market. A particular style that is in trend is dependent on what is popular in society at that time though the style follows a cyclic pattern from the moment of inception till the complete obsolescence from the market.

(a) Introduction -The first stage of fashion cycle implies launching of a new, unique style,different from the existing ones. Most newly introduced styles are available at exorbitant prices and thus are intended for upmarket segment. Fashion leaders, particularly celebrities, rock stars, politicians are the first ones to accept the new styles. Infact, the fashion innovators – designers target this elite social class while launching the new styles into the market as fashion leaders serve as style icons for general masses and can readily buy the latest trends without any price consideration.

Designers are regularlyoffered financial backing and have complete freedom and flexibility to design and showcase their creativity by experimenting with quality of raw materials, or amount of fine workmanshipchanging elements like line, shape, color, etc. This stage ensues that styles are made in fewer quantities, endowing designer more freedom, flexibility and room for creativity. Designers aim at customizing or tailoring for celebrity or role model, for instance a particular mix and match look of apparel and accessory designed for a rock star, politician or Bollywood personality. Production costs are high and are within the reach of elite, high income group people particularly fashion leaders. Designer's new and unique styles are presented to retail buyers and press, as a part of the marketing strategy.

(b) Increase in popularity - As the new fashion is purchased, worn and appreciated by masses, the style begins to rise in popularity. The most expensive assortment of apparels and accessories will never witness very high sales; however, the itemsmay enjoy popularity and likelihood of acceptance as the prices drop. Thus, these styles can turn out to be the most popular of all the high-priced new fashionsin designer's collection. The popularity of the style may further be boosted through copying and adaptation. Keen analysis of designers' collection for the most sort out for styles serve as a guide for manufacturers to buy the rights to produce line-for-line copies (exact duplicate) of the original style, which can thereby be sold at considerably reduced prices. High-priced designers now have secondary sales line accessible to masses at reasonable prices as a result of adoption of original designs.

Other manufacturers may attempt to copy the stylecompromising with raw materials and design details resulting in knock off of the innovative, new concept and design of designer. Other stylists or designers may modify the original designin accordance with the needs of their target consumers. Mass acceptance ensues volume production of the styles initially launched in menial quantities and at unaffordable prices.

(c) Peak of popularity -The consumers of varying financial stature are aware of the newly introduced styles through widespread publicity, endorsements and marketing of the apparels and accessories that are trending. Foot traffic at the retail outlets, willingness to buy the new styles and the ever-increasing demand as prices drop down, prompts manufacturers to flood the market with several knock offs of the original design at many price-levels. This stage of fashion cycle is marked by the availability of the style in range of materials and prices providing lot many options for fashion conscious consumers to choose from, and at reasonable prices. The ripped jean and polo shirt worn by a celebrity in a movie may be the latest rage with every other youngster not just wishing but actually buying it as the style is at peak of its popularity and is available in different price ranges.

(d) Decline in popularity -The peak of popularity for any style of apparels and accessories is followed by the decline in popularity of that style as a result of mass production of many design copies. The style becomes unexciting and easily accessible to masses so the consumers now seek for some unique and refreshing styles in their wardrobe. Therefore, the style begins loosing

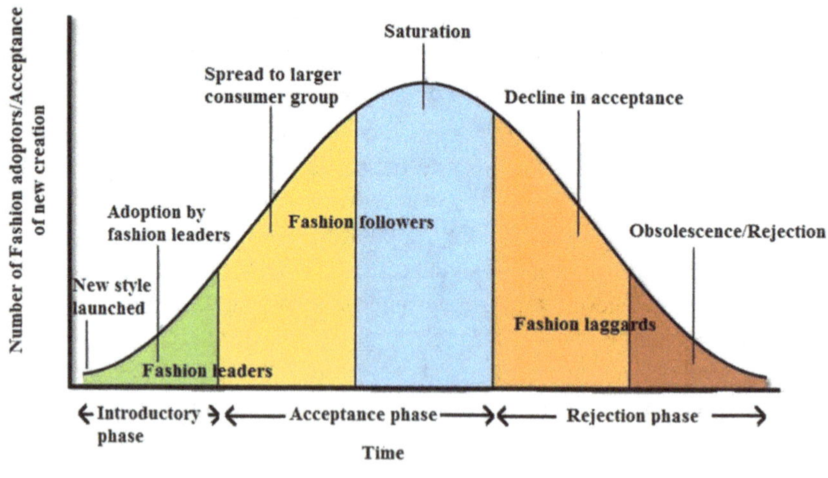

Figure 1.6 Different phases of fashion cycle

popularity and is in its declining phase as consumers are not ready to buy the style at regular prices. Having sensed that the taste and preference of people is changing, retailers prudently decide to clear the declining styles off the shelves by offering discounts and promotional offers, to make space for the new merchandise.

(e) Rejection of a style -Rejection of a style as it is no more likeable and in trend, is called consumers obsolescence. The rejection phase of fashion cycle signifies the inception of yet another style of apparel or accessory which will again follow the same cyclic pattern. The style once disappearing from the fashion arena and completely rejected by consumers may again make a huge comeback with some modifications in style lines, trimmings, raw materials and design details.

1.2.1 Consumer identification with different phases of fashion cycle

The fate of newly launched style of any apparel or accessory is solely dependent on its acceptance by the masses. Appealing styles which fulfill the social, economic and geographical requirements of the population are readily accepted and are the ones that are long to stay and adored by the consumers. Accordingly, the designers and manufacturers target to reach the general masses for wide spread acceptance and in turn mass production of the introduced style. Approval and acceptance by upmarket class which constitutes a small percentage of total population does not guarantee the likelihood of style coming in trend. The target consumers can be divided into different classes based on in which phase of fashion cycle they readily accept and buy the style(Figure 1.6). Consumers can be classified into fashion innovators, leaders, followers, laggards and victims based on their designing abilities, spending capacity and promptness towards acceptance of any new style. The following section discusses the consumer identification with different phases of fashion cycle:

1.2.1.1 Fashion leaders

The small segment of society who quest for new trends in apparels and accessories and wear it before it becomes acceptable by masses are often referred to as fashion leaders. Fashion leaders dare to be different, are believed to have an aesthetic eye, with abundance of money to spend on exorbitant entities and are confident of their own taste. Fashion leaders are the ones to buy and wearnew styles at the introductory phase of fashion cycle. Fashion leaders while travelling across the globe and socialize are ardently searching for unique, latest styles, colors, fabrics, prints, trimmings and ways to accessorize their wardrobe. They are discerning shoppers who usually prefer few exclusive styles in their closet rather than many mediocre, readily available styles. Although fashion leaders are not usually the creators however they deploy a stylist for themselves and in so doing serve as an impetus to a fashion by discovering and wearing a certain style. Designers enjoy experimenting and giving wings to their creativity by designing customized attires for fashionleaders who are individualists, not seeking approval of their dressing style from any one else and do not need the security of standardization. Fashion leaders are considered as style icons for common masses. They are thought of transmitting a particular look by first adopting it and then communicating it among masses.Prominent Bollywood and Hollywood celebrities and movie and TV stars, super models, rock stars, sports persons, athletes, politicians and heirs to royal legaciesare looked upon and emulated as trend setters by common masses. Princess Diana is still remembered for her classy and elegant selection of apparels and accessories.The press reports details, media coverage of lifestyles of these celebrated, distinguished public figures propagates massively about their dressing, embellishments and accessories at public events, in films, or on television. Aishwarya Roy, Priyanka Chopra from Bollywood industry become talk of the town as far as their wardrobe and accessory selection while walking on red carpet for Cannes film festival is concerned.It can thus be pointed out that fashion leaders serve as stimulant to designers and to the fashion industry as a whole owing to their backingto couture and boutique designers over the years, who in turn have flexibility to design the most exquisite collections without any compromise on creativity and expenses. They dare to stand out distinctly from the masses and their acceptance of a designer's label or clothesline makes the clothes fashionable and trendy.

Categories of fashion leaders

Fashion leaders fall into two categories: fashion innovators and fashion role models. Fashion innovators are the celebrity designers, stylists who use their creativity and designing prowess to come up with exclusive collection of apparels and accessories. They primarily prefer designing customized high end products targeting high income, elite group of society as this provides them with ample opportunities to experiment with range of materials, colors, details and silhouettes without any financial constraints. Innovators can take inspiration from several sources like nature's beauty, flaura and fauna, geometrical shapes and patterns, street fashion. However, sometimes designers tend to imitate other famous designers clothing lines and labels to come up with knock offs or first copies of designer's original collection compromising materials, trimmings and design details. Knock offs serve as cheap substitute for middle income fashion followers who tend to follow style icons but wait for the style to be affordable, ignoring the variation in materials and design details of the original style.

Role models also referred to as Avant garde or fashion forward are not necessarily creators of the fashion or are not the first ones to wear the latest styles. They try to find unique fashion in small boutiques or vintage clothing stores, or hire designers, stylists to design apparels and complimentary accessories for them. The newly introduced, exclusive style creations presented by innovators are endorsed by role models in their movies, campaigns, page 3 parties and are photographed by the press so on. The adoption of look byearly adopters i.e. role models has positive influence on the flow or distribution of fashion. Role models, thus serve as the visual communicator of the style with innovators using them as media to promote and publicize their creations (Figure 1.7).

Figure 1.7 Fashion leaders – Role models

1.2.1.2 Fashion followers

Fashion need followers, or it would not exist. Any newly introduced style will be adorned, last long and will be in trend only when acceptable by masses rather than classes. Mass acceptance of any newly introduced style is very crucial to motivate the manufacturers to manufacture the styles in bulk and at affordable prices. Fashion followers closely follow public figures - world, national or community fashion leaders to be confident of the styles they ultimately choose for themselves and seek acceptance through conformity. Copyists or adapters produce cheaper, imitated versions of the original concept targeting fashion followers who will readily accept the imitated version without giving much consideration to material or any deviation from the original design.

Profitable volume mass production of the exclusive, fewer quantities of initially designed styles is made possible by increasing demand and acceptance by fashion followers. Fashion followers constitute 68% of total population and are held responsible for making any particular trend or look fashionable (Figure 1.8). 34% of the fashion followers constitute early majority who accept and wear the merchandise as the style is advancing towards peak of popularity and available in retail outlets at reasonable prices.The remaining 34% belongs to late majority who wish to be stylish as well and follow the role models yet accept the trend when it has heavily flooded the market and different versions of the styles at varying prices are readily available. Consumers consider it worthwhile to be fashion followers rather than fashion forward for several reasons:

- Lack of time, money, resources and willingness forsincere devotion towards fashion leadership
- Need a period of exposure and security of standardization to new styles before accepting them
- Insecure about their tastes and therefore look up to fashion leaders and role models for acceptable and appropriate styles
- Have an urge to be part and parcel of their peer group
- Tendency to imitate and impersonate role models whom they admire

1.2.1.3 Fashion laggards

Fashion laggards constitute 16%of the total population (Figure 1.8). This consumer group is negligent and indifferent towards the styles that are trending or hot selling products. Fashion laggards do not pay any heed to stay updated or follow role models for new styles. They are lagging behind in fashion and adopt the style as it is at the declining phase of fashion cycle. Lack of resources, finances and interest towards fashion may be some of the factors which prompt consumers to follow trend as it is declining. Fashion laggards generally prefer buying the styles during promotional offers like end of season sale irrespective of unavailability of color and size options.

1.2.1.4 Fashion victims

This consumer group has means and resources to buy expensive apparels and accessories. However, they lack the foresightedness and wisdom to decide and choose wardrobe that flatters their body types. Consequently, they will buy and wear styles that are in trend without any consideration towards their age, body types, circumstances, physical traits like complexion, height etc. and in so doing become slaves to designer brands. Fashion victims blindly and stupidly follow a brand without any discernment and analysis. Fashion victims unwittingly become mannequins as they mindlessly follow the dictates of the fashion world. Some characteristics common to fashion victims are:

- Wear unflattering trends and do not choose apparels and accessories in accordance with their body types.
- Tend to go overboard on a trendand aim at displaying all the accessories and embellishments in one go be it jewelry, makeup, clothes, or perfume.
- Fashion victims have a tendency to wear has been trend. They spend huge amounts on their clothing thus, are not convinced to bid farewell to obsolete styles soon and continue wearing a style when it has virtually disappeared from the social arena. The disco era left fashion victims so much mesmerized that they continued wearing polyester suits and several chain necklaces for years at a stretch after the trend had faded.
- Following trends regardless of weather, health, surroundings and age – fashion victims prioritize to show off their fashion sense and look good that very often suspend their common sense. Latest fall or spring line in the stores may instantly prompt the fashion lemmings to grab and wear it regardless of the weather. Likewise, fashion victims often invite unprecedented troubles for themselves by wearing trends to places that can jeopardize their health or safety e.g. high heels in snow clad mountains, synthetic loose clothes and lot of embellishments to fire prone regions.

1.3 Classification of styles

Fashion is observed to follow the same cyclic pattern. However, there is no measurable time table for a fashion cycle. Some styles reach the peak of popularity promptly as they are introduced. Nevertheless other styles are slow moving which take long duration to enjoy wide spread acceptance and popularity.Certain styles of apparels and accessories, namely fads lack the design strength to hold consumers attention for long and witness decline in popularity very rapidly, still other styles are instant hits busting to popularity and then to decline, both rapidly as shown in Figure 1.9. Some styles are seasonal and will sell in a single season when they experience mixed response from consumers and will thereafter completely be eliminated from retail outlets while others repeat themselves for several seasons at a stretch. Styleswitness their glorified state or dismal based on the era, social and economic preferences of individuals and Nation as a whole. Womens'ethnic attire accessorized by turbans was trending in the 19th century, but is rarely worn today except as part of particular religious beliefs and social norms like in Sikh culture, weddings and processions. Styles are classified as classics, fads, recurring cycle and interrupted cycle based on the length of fashion cycle the style spans, design strength, acceptance, popularity and ability to hold consumer's attention and sustain in the market for long.

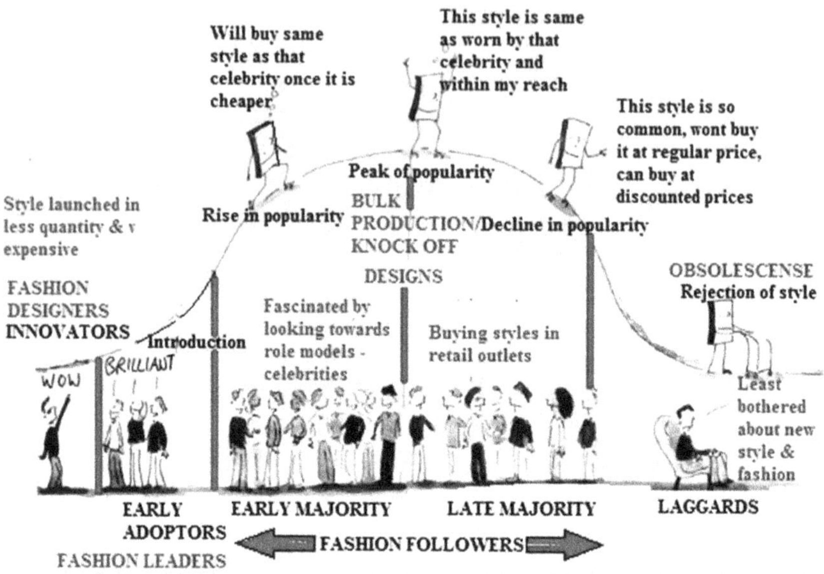

Figure 1.8 Consumer identification with different phases of Fashion Cycle

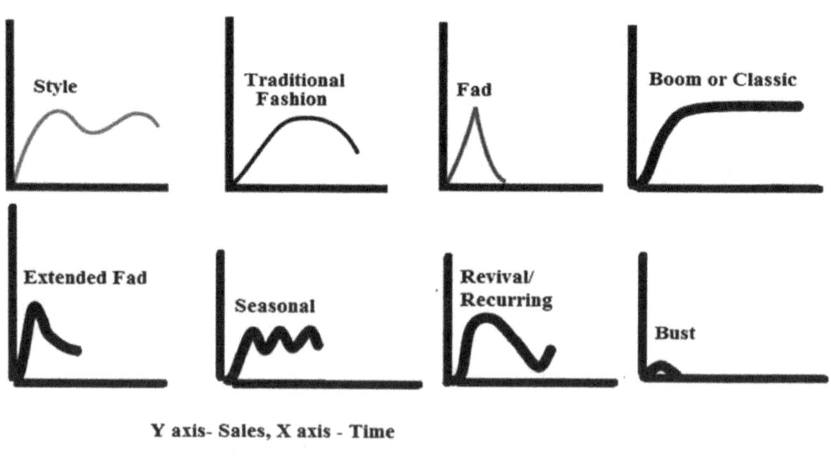

Figure 1.9 Comparison of different styles

1.3.1 Classic

A classic fashion is a style that remains acceptable and enjoys popularity among consumers of various economic strata for several seasons, from one year to the next and sometimes even several years. Classics are also referred to as ageless styles as they never become completely obsolete owing to their long lasting and wide spread acceptance among different consumer groups. The salient feature of classic styles which prompt masses to adopt them and not getting bored of them is the simplicity and design strength which is successful in drawing and holding consumers attention for quite long. Blue jeans are a striking example of classic style which emerged as a staple of dressing in the United States in the twentieth century. The style lines and design details may change from one season to the next, including stone washing, dyeing, painting, tearing, and fraying although the basic design of jeans remains the same. Blue jeans epitomize the growth of casual fashion and their endurance is ensued because of resonance with changing eras and consumer's likings and preferences.Sneakers – flat running shoes, ballerinas, baseball cap, saree, formal shirt, a worsted wool suit, a leather bomber jacket, A-line kurta, black corporate tailored suits, polo shirt, lehenga set are other examples of classic fashions that enjoy repetition year after year (Figure 1.10).

1.3.2 Fad

Fad is a short lived fashion that lacks design strength to hold consumers interest and liking for long duration and usually lasts only one season, at times even less than a season. A fad tend to be at the extreme end of a design and never quite makes it into the mainstream owing to complexities in style lines, design details, selection of colors, motifs, trims and notions so that general masses do not relate well and hesitate to accept it. Fashion followers may adopt the fad trend readily thrilled by the endorsement of role models, however will soon be dejected by the style and quest for a better, persistent option. Consequently, fads appeal to smaller section of society who wishes to experiment with their attires and so these styles have a short stay in market. Figure 1.11 shows the different fad styles which instantly hit the market and then promptly become rejected and unacceptable as well. Sequined jackets and tops, embroided jeans and footwear, fringed shrugs and scarves, tasseled earrings, ripped jeans, platform shoes, printed stilettoes,leather pants, paper dresses, belly and eye piercing, tattoos, spikes and highlighted hair, umbrella sleeves, tie and die tops and t shirts, frilled tops, peculiar and eye catchy fluorescent, neon colored dresses, t-shirts are some examples of fad fashion.

Figure 1.10 Classic styles

Figure 1.11 Fad styles

1.3.3 Recurring styles

New, unique styles are introduced, follow a cyclic pattern as dictated by fashion life cycle, and ultimately get disapproved and rejected. However, some styles of apparels and accessories are revived and repeat themselves after a lapse of several years. Therefore, any style after its decline and rejection phase may resurface many years later with reinterpretation of fabrics, colors, textures, prints, motifs, style lines and design details. Designers take their inspiration from nostalgic looks of the last century keeping the style, silhouette or proportion intact although presenting the past trend in completely transformed, new form to appeal today's fashion forward consumer.

Figure 1.12 Recurring styles

Accordingly, nothing is ever exactly the same, yet nothing is totally new. Comeback of the styles may be huge hit among the masses but occasionally the trend can receive a mixed response as the consumer may not always be ready to invest in a look that they consider does not fit the criteria of present times. Style that does not make a comeback may be recreated. The recurring styles exhibit contrasting differences from classic and fad styles owing to their complete dormancy and non-existence in the market for indefinite period of time referred to as latent period (Figure 1.12 & 1.13) and thereafter revival and comeback with a fresh, appealing outlook. Jeans have enjoyed the status of staple merchandise that is acceptable by consumers irrespective of their financial status, physique and age. Designers, manufacturers and retailers try exploiting inclination of consumers towards this lower torso apparel by presenting jeans in new form year after year like skin fit, boot cut, low waist, high waist, different washes, torn and distressed jeans. Leggings are another example of recurring style that made a comeback with major changes in raw material and design details. Leggings werehot selling articles outselling jeans in the United States by the early 1990s, however they declined in popularity with the introduction of wide pant legs in the late 1990s. Leggings witnessed revival in 2005 with basic silhouette remaining the same but with variation in materials – lyre, spandex based, printed, shiny, metallic, leather and suede. Chokers, palazzo pants, crop tops, clutch handbags, hair styles like braids, buns, polka dot fabrics, long kurtas and skirts, anarkali suits, cold shoulder tops, ponchos, kaftans, shimmery pants, bell bottoms (Figure) are several examples of recurring styles that have garnered come back with changes in one or other detailing.

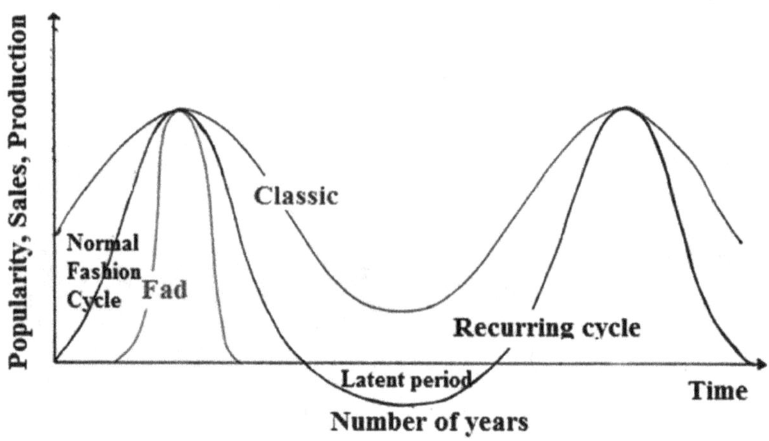

Figure 1.13 Comparison of classic, fads & recurring cycle

1.3.4 Interrupted cycle

The normal progress of fashion cycle may sometimes be affected by several extrinsic, uncontrollable social, political, economic factors like world war, emergency, terrorist attacks, natural calamities, market analysis and changing preferences of consumers. All such circumstances may result in interruption or prolonging of the existing style. Women's fashion in the 1930s constituted of large- shouldered, wedge shaped silhouette and the same silhouette continued without expected decline and rejection during World War II. Fashion took a back seat for people in such difficult times as they were struggling for their livelihood and survival. Unnatural prolonging of the old style brought about a radical change with the introduction of sloping shoulders, tiny waists, and longer skirts as world peace was restored in 1947. Likewise, if the country's economy is adversely affected, the prime focus will be social welfare and fashion will be a thing of secondary concern.

The production of a particular style may be prematurely halted by manufacturers due to decline in popularity of that style. Manufacturers and retailers become wary of style not being accepted by consumers at marked prices and consider it worthwhile to discontinue production and stocking of merchandise that will soon decline in popularity. Consumers attempting to buy winter clothes in April would end up returning empty handed as the products would have been taken off the shelf as summers approaches.

1.4 Fashion theories

Fashion theories are crucial part of fashion world in providing an insight on the propagation and adaptation of fashion as per the taste, life-style and budgets of various costumers.Fashion adaptation process encompasses three variations- traditional adaptation, reverse adaptation and mass dissemination. The origin, outlook, extent of influence and motivation of adopters determines the manner in which newly introduced style or trend permeates among the individuals. The distribution and likelihood of acceptance of fashion has been described as a movement, a flow, or trickle from one class of society to another. The movement of fashion may occur downwards from fashion leaders and role models to fashion followers, upwards from fashion followers to role models, celebrities and lastly across consumers of different social strata simultaneously as shown in Figure 1.14.

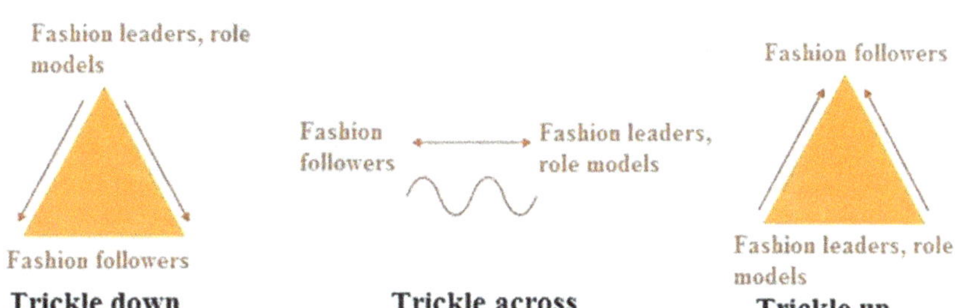

Figure 1.14 Fashion adoption theories

1.4.1 Trickle down (traditional adaptation)

Veblen described the oldest theory of fashion distribution as the trickle-down theory which is considered to be traditional adaptation. According to Trickle-down movement, the style exhibits downwardmobility and shows dependency upon a hierarchical society. In this model, a style is primarily offered and adopted by upmarket, elite group of people in society i.e. fashion leaders and role models and gradually progresses andbecome acceptable by middle income group people i.e. fashion followers (Figure 1.15). Couture designer fashion and traditional process of copying and adapting trendsetting fashion from, Paris, London, New York marks the beginning of this stage. The style initially customized or tailored for celebrities is expensive andaffordable by elite class only. Designs are eventually copied again and again at lower prices until they have been endorsed, publicized for conservative buyersand accepted by them.

Figure 1.15 Styles trickled down from celebrities, role models to common masses

This distribution model assumes a social hierarchy in which common masses with just adequate means tend to identify and imitate the well off, influential people, who in turn quest for distinction and uniqueness. As the style becomes wide spread and common among fashion followers, affluent, wealthy class rejects that look for another new innovative concept. Pillbox hat, pearl accessories, fur coats and sweaters are examples of apparels and accessories which were first adopted by Richie rich, elite class and then later on followed and accepted by middle and lower classes of society.

1.4.2 Trickle up theory (reverse adaptation)

The trickle-up theory also known as bubble-up pattern is initiated from the streetand adopted by high income group. The movement and adoption follows reverse adaptation with the trendinitiating from street fashion and approaching towards high fashion.Designers take inspiration by closely following and analyzing youth, general masses who may follow mix and match style of apparels and accessories. The street or youth culture may switch over to a new look as thestyle isadopted by affluent, upmarket, older or more traditional consumers. This newest fashion adoption theory reflects changing styles and practices in fashion.Ripped jeans, leather jackets, gypsy look are some examples of a street look which made it big to the runway (Figure 1.16). Likewise, T-shirt and blue denims initially worn by laborers and working classes as a functional garment havebeen adopted universally as a casual outer wear and essential components of the everyday wardrobe.

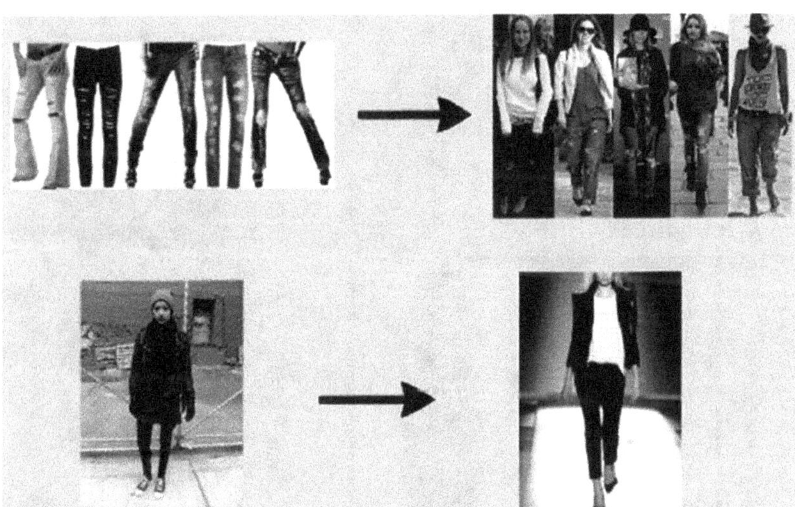

Figure 1.16 Styles trickled up from teenagers, common masses to affluent, upper class

1.4.3 Trickle-across theory (mass dissemination)

Proponents of the trickle-across theory claim that fashion moves horizontally between different consumer groups irrespective of their economic status. Trickle-across model is mass dissemination theory of fashion adoption where there is little lag time betweenadoption from one group to another. This theory is supported by the fact that designers present particular look of apparel or accessory simultaneously at affordable prices for all sections of society. Accordingly, the styles do not follow the conventional trickle down movement howeverwill appear in range of prices at approximately the same time. Designers explore variations in similar styles to meet the requirements and aspirations of all the consumers alike. As the design makes it to the runway, manufacturers attempt manufacturing a variety of knock offs of the original style with compromises in selection of materials, trims and notion. Widespread access to fashionis encouraged by mass production as many different styles at variety of price pointsare available. A particular style of apparel and accessory that is trending is made available in discount, department and designer stores at approximately the same time.Rapid mass and digital communications, promotional and endorsement endeavors of manufacturers and retailers, and exposure of a look to fashion leaders encourages the adoption of introduced style among various economic classes of society concurrently.Tote handbags and sheer dresses are other examples of trickle across theory (Figure 1.17).

Figure 1.17 Styles trickled across among various consumer groups

2

Fashion Accessories – Introduction & Classification

2.1 Introduction to fashion accessories

Fashion accessories are the additional entities which compliment, supplement and enhance the overall look of the garment. Fashion accessories add color, style and class to an outfit and go a long way in highlighting the aesthetics and appeal of wardrobe in a secondary manner. They are also considered visual embodiments of religious affiliation, an individual's identity and disposition. Accordingly, fashion forward and conscious consumer always chooses an accessory be it functional as handbag, watch, footwear, head covering to decorative ones like jewelry, hair accessory with utmost care and wisdom. Just like apparels, accessories can be long-lived, timeless classics, that remain in fashion for year-after-year like jewelry, time wear or can be trendy, fad items that come and go in single seasons like funky hair accessories, floral scarves, etc.

The accessories are often chosen with the aim to complete an outfit as well as complement the wearer's look. Although the concept of accessories in India has been borrowed from the west, the industry has great potential in national market witnessing revolutionary growth over the past few decades on account of the ever increasing demand and preferences of accessories by consumers irrespective of their gender and age groups. The designers and producers of accessories therefore, must be aware of ever changing needs and aspirations of today's fashion conscious consumer who seek value for money. Moreover, there has been implications and greater challenges on accessory manufacturers as they have to keep up pace with fashion trends prevailing in apparels to captivate consumers and complete the look.

Accessory designers and manufacturers are working in sync with fashion forecasters, merchandisers and focusing on market trends and latest rage in apparel industry in order to fulfil the functional as well as aesthetic expectations of accessory users and buyers. Accessories like hats, footwear although being functional are being designed as more and more aesthetically appealing with surface ornamentations and use of different sundries. A hat is a head covering which can cater to different applications and can be both decorative as well as functional. In the middle ages, hats for women ranged from simple scarves to elaborate hennin which were an indication of social status and financial well-being. A wide variety of hat styles are popular now-a-days like fedora, bowler and cloche hats. Footwear are necessary components of wardrobe which not only provide an effective covering to feet and protection against rough ground and extreme climate like thongs, mules styles in females and wingtip, moccasin styles in menswear but also play a significant role in enhancing an individual's personality by giving illusion of being tall like in case of high heel footwear stilettoes, pumps or a royal look by decoratively embellished clogs, espadrilles. The importance of fashion accessories can thus not be undermined as this industry contributes equally to the market profitability as the apparel industry and holds a greater potential of expansion in national and international fashion arena.

2.1.1 Importance of accessories in apparel industry

The Indian apparel and accessory market has grown manifold over the past few years and is swarming with a vast range and variety of products offerings like apparel accessories - stoles, scarves and hard accessories like bags, fashion jewelry, time wear and eyewear. The accessory industry cannot be underestimated with an overall contribution of 10% to the fashion industry. Accessories are no longer add-ons, but have been successful in maintaining valuable identity in fashion industry. Accessories symbolize expression of an individual's personality, tastes, preferences and sense of style and are merely not considered need-based articles. Accessories lately enjoy status of becoming mainstream entities rather than just a category to complement apparel. Fashion is no longer associated with clothing items, but gone beyond that from a pair of kurta and churidaar to the right match of jewelry, hair accessories, sun glasses and footwear. Similarly, for men – a pair of denims with formal white shirt teamed with leather belt, watch and sneakers add a style statement. The accessories thus play a pivotal role in shaping the journey of the consumer through the social strata in ever changing fashion and brand consciousness scenario.

Accessories are equally important as clothing and offer a completely whole new opportunity to express yourself. Accessories and clothing go hand-in-hand and possess equal significance in an outfit; working together to create an ensemble which is an expression of style and individuality. Clothing may physically take up more space in an individual's closet and outfit, nevertheless accessories are important details needed to render final, finishing touch to the look. A retro look is incomplete without the complimenting braided hair do or a high positioned bun, pearl teardrop earrings, a necklace and broad framed colored sunglasses. A handbag, neckpiece, scarf or wristband may not make much of an impact alone but correct mix and match of any of these accessories with apparels render outfit a visually appealing impression. Evolution of Indian consumers and their fashion sensitives has accelerated the demand of wearable as well as drape-able accessories and led the industry in boom mode. Urban India witnesses major drift and liking towards fashion accessories irrespective of gender and economic status of the consumers. Its interesting to note that for fashionable women its not just their clothes which look great, rather they are always punctuated with great accessories.The industry has enormous potential for growth and expansion as the accessory market is presently dominated by unbranded players and unorganized sector, thus a very large part of the accessory market still remains untapped by the organized sector.

2.2 Categories of fashion accessories

Fashion accessories can be categorized on the basis of a number of factors:

- **Raw material usage** – textile and non-textile, trims and notions - a wide variety of raw materials, trims and notions are used for accessory designing. Accessories can be classified as those designed and constructed from textile materials and the others using non-textile materials – metals, gemstones, leather, fur, suede. Metallic buttons, snap fasteners, polyamide velcro tapes, elastic are some of the trims and notions used in accessory-making.
- **Construction methods** – the construction methods of fashion accessories vary as per accessory to be designed, i.e. fabrication techniques like knitting, weaving are used for textile-based wearable accessories like gloves, socks, stockings, mittens, etc. Lost cast waxing technique, embossing, annealing, soldering techniques are used for metal-based fashion accessories like fashion and costume jewelry.
- **Styles** – fashion accessories can further be classified based on styles. Footwear are available in different styles based on foot exposure and type of heels like thongs, sneakers, kitten heels, stilettoes. Clutch, sling bags, totes, envelopes are some popular handbag styles among women.
- **Utility/functionality** – Another classification of fashion accessories can be based on their functionality. Accessories like eyewear, handbags are functional and generally used in day-to-day routine, the former providing better vision and protection to eyes against

scorching sunlight while the latter are used for travelling, carrying necessary articles. Decorative fashion accessories like jewelry do not serve any particular function rather they adorn an individual's appearance and add style quotient.

- **Styles of wearing - draping and grasping/hand held** - Fashion accessories can further be categorized based on whether they are wearable like footwear, cravats, ties, hats, bonnets, elts and suspenders, gloves, muffs, jewelry, watches, sashes, shawls, scarves, socks, stockings, jewelry, and eyewear or those that are grasped or handheld like clutch handbags, parasols, umbrellas, fans, ceremonial swords, etc.
- **Enhancers and Statement creators** - another classification can be on the basis of style quotient - Enhancer and Statement creator. Enhancers ensure that the consumer's status in peer group and society is enhanced or uplifted like belts, handbags, wallets, socks. Females are generally seen flaunting branded handbags, eyewear. Likewise, men are found boasting of Rolex or Rado watch which serve as an enhancer for their status and personality. Statement refers to categories of accessories that create a style statement like scarves, headgear, ties, key chains, tie-pins and cuff-links which are generally affordable with style and design details changing from season-to-season.

The detailed description of raw materials utilized for accessories and the different styles of each category will be discussed in detail in the subsequent chapters.

Table 2.1 Categories of fashion accessories

Category of fashion accessories	
Handbags	Handbags, Totes, Backpacks, Duffle Bags, Sport Equipment Bags, Messenger Bags, Briefcases/Attaches, Laptop Bags, Fanny/Waist Packs, Diaper Bags, Luggage Suitcases, Garment Bags, Luggage Sets, Travel Toiletry Cases, Wallets, Change Purses, Cosmetic Bags
Time wear	Casual Watches, Dress Watches, Sport Watches, Body Jewelry Watches
Eyewear/sunglasses	Fashion/Everyday Sunglasses, Sport Sunglasses, Sport Goggles
Women's & Men's footwear	Pumps, stilettoes, Ballerinas, Espadilles, Sneakers, Moccasin, Wingtip, Oxford
Jewelry	Earings, Rings, Bracelets, Anklets, Chains, Neklaces, Pins, Brooches, Jewelry sets, Cufflinks

Figure 2.1 Categories of fashion accessories

2.2.1 Functional accessories

2.2.1.1 Footwear

This category of fashion accessories serve functional purpose providing protection to wearer's feet against extreme climate and abrasion against rough, pebbly surface. Comfort is although given precedence while designing and manufacturing of men as well women footwear, however, aesthetics is becoming equally important these days with consumers matching and complementing footwear with their attire. Women prefer pairing shoes with trousers or skirts and handbags, while gentlemen compliment footwear with their belt's color. Shoes, sandals and boot is the largest category of accessories catering to both dress and casual trends, with more than seven billion pairs of shoes produced and sold worldwide each year. Footwear are manufactured in assorted materials including canvas, jacquard woven fabrics, denims, calf, kid, suede, and reptile skins, imitation leathers. Several footwear styles as per feet coverage area, height of heels and grasping elements are trending. Thigh-high, knee thigh, cowboy and wedge booties are some of the innemorus options available for boot lovers while the consumers having penchant for flats can choose from among ballerinas, flip-flops, slip-on, dockside, flat-sandals and gladiators. Footwear intended to give illusion of being tall include high heel footwear like peep toe, kitten heel, stilettoes and pumps.

2.2.1.2 Hats

A hat is a head covering manufactured in variety of materials and styles based on the purpose and end-use requirements of the wearer. The most common use of hats is in providing protection against ambient conditions like humidity, sunlight, rain and severe cold. Sun hat with a wide crown and brim covering the head and forehead completely is preferred for summers to mask the effect of scorching sunlight while beanies, skull caps snugging tightly over the head provide warmth in winters. Likewise, fur hat with fold-down earflaps covering head and ears completely is a basic necessity in tropical and snow covered regions. Besides protection against environmental conditions, protection to head is often required at workplace to ensure safe work environment and accordingly head gears and complete head covers, caps are recommended to overcome any occupational hazards. Hard, metallic hat protects construction workers' heads from injury against falling objects; custodian helmet protects the traffic police men head. Hats can also serve as an indicator of social status, profession like toque worn by chefs, peaked caps or brimmed hats worn by the Royal Canadian Mounted Police. Denotation of nationality, branch of service, rank regiment, e.g. in military also demands wearing a specific type of hat.

Moreover, the hats hold religious significance with individuals of particular communities covering head and wearing hats as religious, cultural obligation like the mitres worn by Bishops and the turban worn by Sikhs. Furthermore, hats are also popular for ceremonial occasions like mortarboard worn during university graduation ceremonies.

2.2.1.3 Handbags

Are generally paired with a pair of shoes the female is wearing. Women prefer handbags for carrying all their beauty essentials. Wallets, purses and card holders carried by males are complimented with color of trousers and belts. Handbags render a fashionable look when hanged around the hand, at the arm or hand arm or wrist base. Handbags are designed in all trendy colors, fabrics, and variety of shapes to compliment the wardrobe. They can be decorative as well as functional; holding articles and necessities conveniently and at the same time fitting into the fashion picture. Bags such as totes, satchels, portfolios, or backpacks are spacious and larger in size carried for travelling or carrying substantial amount of belongings and are functional while smaller bags such as clutches or envelopes are usually decorative and complimented with the attire for different occasions. Handbag styles range from classic, constructed types to soft shapes. Leather, including suede and reptile, still represents approximately half of handbag material; vinyl, fabric (tapestry, rug prints, needlepoint, silk, wool, nylon and canvas), and straw make up the other half. Major trends include a wide variety of shapes in an array of colors, some with beading or embroidery.

2.2.1.4 Watches

Are considered to be the must have accessory for consumers, irrespective of gender and age group. The watch industry does not remain unaffected by constant urge of consumers seeking functionality and aesthetic appeal simultaneously in any entity they wear. Accordingly, with the changing times and preferences of watch users, watches have witnessed quite a drift from conventional analog watches to the state of the art luxury watches embellished with precious gemstones to smart watches which can track and monitor intensity of sports person's physical activity. Components of watches are assembled in such a way that displays meticulous workmanship, functionality and design. The watch making process starts with creation of rough sketch on paper, which defines the proportions and aesthetical elements. Post the sketch making; the movement is designed around the concept which takes into consideration both functional and aesthetic aspects. Lastly, a prototype is manufactured (Figure 2.2) which is functional yet raw version of both the movement and the case in order to test the watch's working before mass production. A variety of materials which are key design elements of a watch are used for watch construction - materials; the case, band, strap, crystal and movement. A lot of consideration is thus required while selecting materials for different components of watch as they may have a direct bearing on the durability, longevity or performance of the watch.

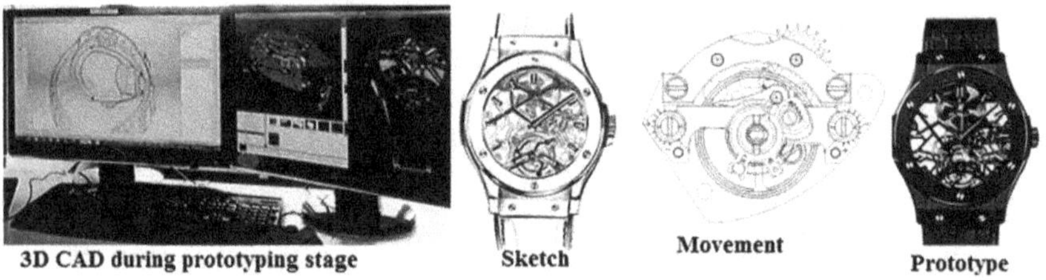

Figure 2.2 Protyping design & development for Timewear

2.2.1.5 *Eyewear*

Eyewear constitutes the functional wearbale category of fashion accessories which safeguard the eyes from excessive sunlight. Eyes, being light sensitive organs, require protection from overexposure to radiation in the visible and non-visible spectra. Women's eyeglasses are now available in playful cat-eye shapes, as well as cutouts with elaborate design elements including swirls and angular patterns. Men's eyeglasses include classy designer logos and patterns embedded in translucent backgrounds. Unisex eyeglasses for both men and women include plenty of rimless styles, which are popular as a good, generic fit for both sexes. Kids› eyeglasses may look similar in style to those worn by adults, but are often more colorful with emphasis on durability. Sunglasses make a fashion statement, they also serve as an important device to protect your eye›s health from damaging UV rays. Designer sunglasses styles, often branded with popular names such as Armani, are generally sold in optical boutiques or departmental stores and tend to mimic current fashion trends. Prescription sunglasses are ideal for prescription eyeglass wearers and eliminate the need for clip-on sunglasses. The rims lend form and character to eyeglasses—they also provide function by holding the lenses in place. The end pieces are the small parts on the frame that extend outward and connect the lenses to the temples. The bridge is the center of the frame that rests on nose and joins the two rims together. The hinges, which sit between the end pieces and the temples, allow you to close your glasses by folding the temples inward. Lenses are the clear pieces of glass, plastic or other material held in place by the rims. The lenses are crafted and shaped with your unique prescription to help you see clearly. The screws are the small metal fasteners near the hinges that connect the end pieces with the temples. The nose pads are the round plastic pieces under the bridge that sit on your nose. They give your glasses a more comfortable and secure fit. The pad arms extend from the rims and hold the nose pads in place. They are adjustable to fit the natural shape of your face.The temples are the long arms on the side of the frame that fit over your ears for a snug fit.

Frame styles branded with high profile designer names like Calvin Klein, Jones New York, and Marc Jacobs continue to be in demand. And even for youngsters who might ordinarily dread the «nerd» factor of wearing eyeglasses, popular fiction characters like the bespectacled young wizard, Harry Potter, have become role models.

Retro nerd reading glasses

Glasses with photochromatic lens

Glasses with trendy frames

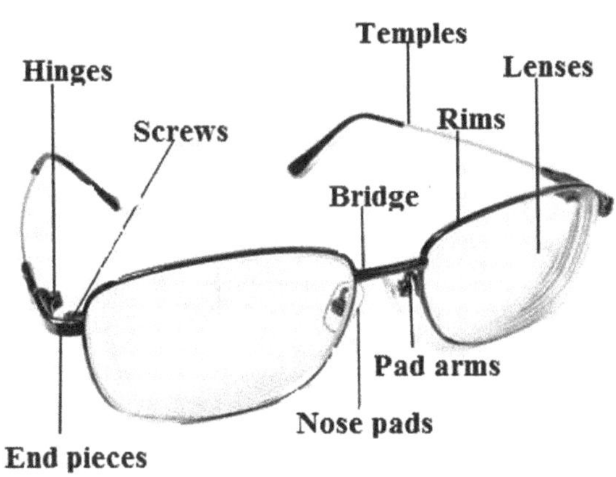

Figure 2.3 Eyewear & associated components

Figure 2.4 Functional fashion accessories

Figure 2.5 Wearable fashion accessories

2.2.1.6 Scarves

Are drapeable accessories which play a vital role in providing protection against external environmental elements and extreme climate and at times draped around head and face to fulfill religious and cultural obligations. Mufflers, head covers, neck cloths, waist ties or bag accessory (a piece of cloth tied to the handle of the bag also constitute this category. However, the ambit of scarves is not just confined to functional aspects per se and can also be draped innovatively to conceal the figure flaws, accentuate and refine the wardrobe. Scarves when draped around women's neck as cowl or rectangular fabric loosely hanging down present visual delight (Figure). Scarves are designed and developed in a range of materials, shapes and sizes. Fabrication method employed for scarves' production determines their texture, appearance and weather-pertinence, and are classified based on shape, color, texture, size, pattern, season and material. Natural as well synthetic fibres namely cotton, silk, pashmina, linen, cashmere, chiffon, alpaca, satin, wool, and acrylic are extensively used for woven and knitted scarves and mufflers. Furthermore, scarf designers and manufacturers experiment with an array of beautiful prints such as geometrics, floral, polka dots, plaids, paisley,animal prints, embroidered or hand painted motifs finished with equally appealing trims.

2.2.1.7 Neck ties

This accessory category augments the appeal and are usually matched with tailored dinner jackets, blazers and formal wear. Men and women adorn neck ties equally which impart clasy, formal look to men while women prefer wearing neck ties to portray sleek look. Neck ties are available in a variety of materials – silk, satin and other synthetic fibres and blends with prints ranging from plaids and polka dots for formal ones to figurative, flaura and fauna motifs to render a casual look (Figure 2.7).

Figure 2.6 Various draping styles of scarves

Figure 2.7 Different print & draping styles of mens' neckties

2.2.2 Knitted accessories

Knitted accessories are versatile, finding utility in summers as well as winters and generally preferred by consumers over woven fabrics and accessories. Summer knitted accessories like driving gloves, wrist and head bands, pantyhoses and stockings are intended to provide good moisture absorption and release properties alongwith reasonable stretchability and accordingly, are developed from fine yarns, hydrophilic fibres and open structures of low aerial density and thickness with extensive use of elastane, spandex and lycra. Likewise, thermal insulation is the most sought after property for winter wear accessories like socks, gloves, scarves, beanies and skull caps, ponchus and mittens. The materials chosen for winter wear accessories comprise of high denier yarn, compact structures of high aerial density and thickness. Non-textile materials like fur, suede are also used for winter wear accessories to impart warmth retention properties.

2.2.3 Sports accessories

The increasing awareness and inclination of population towards health and fitness prompt them to indulge in any physical activity or sports to keep various diseases at bay. People prefer wearing comfortable yet trendy sportswear complemented with appropriate accessories when hitting the gym or indulging in walking, jogging or running. A dry release, quick moisture absorbing t-shirt with a stretchable track pant, base ball cap, a comfortable pair of sneakers, wrist band and head band would complete the sporty, trend and effervescent look of avid fitness freaks. The urge of consumers to stay fit and choosing their attire as per level of physical activity, provides a great impetus for the sportswear apparel and accessory manufacturers to experiment with innovative raw materials, technologies and designing aspects thus developing entities to suit the functional and aesthic needs of consumers. Accordingly, a lot of research and development goes into the designing and engineering of sportswear apparels and accessories. Innovative and high-performance functional fibres, yarn and specially engineered knit structures are making their way into sports arena to provide comfort, stretch, thermal insulation, moisture vapor and liquid moisture transmission properties. The sports accessory industry also follows the suite with a range of sportswear accessories being engineered and developed to keep in pace with the most sought after requirement of comfort, servacibility and aesthetics by sportsperson. The innovations in sportswear accessories are fascinating and multi-dimensional from staple articles like bag packs, gym bags, duffles, base ball caps, visors, diving and underwater glasses, head covers and head gears, biking, tracking gloves, wrist and head bands at one end of spectrum to sneakers with spikes capable of enhancing runner's performance to socks designed with odor absorbing, micro-encapusalted frangrance releasing materials and smart watches capable of monitoring level of physical activity, physiological body parameters like heart beat, pulse rate and calories burnt on the other end of the spectrum.

Figure 2.8 Knitted accessories

Figure 2.9 Sports accessories

2.2.4 Decorative/Ornamental accessories

Decorative/Ornamental accessories – serve as a means of self pride, adornment and wealth display at times. Women usually display their signature look by mix and match of complementing jewelry with wardrobe as per occasion. This category of accessories have limited functionality associated with them and are designed and developed paying greater attention to designing and detailing aspects. Jewelry and hair accessories belong to category of decorative/ornamental accessories.

2.2.4.1 Jewelry

Can be broadly classified into costume jewelry and fashion jewelry. Costume jewelry is unique, classic class of ornaments designed with intricate and fine workmanship using high quality expensive metals like gold, silver, platinum and precious and semi-precious stones like pearls, ruby, sapphire, emerald, etc. This category of accessories is generally preferred for important occasions and ceremonies owing to their inquisitiveness and high cost. Fashion jewelry is fad, trendy imitation of jewelry intended for daily routine and crafted out of less expensive materials like artificial stones, beads, sequins, ivory, leather, various trims and notions. Costume and fashion jewelry includes products like bracelets, necklaces, earrings, bangles, pendants, finger rings, toe rings, anklets and brooches. Some popular earing styles include drop earrings, studs, jhumka, ear cuff, chandelier earrings, danglers, bali, hoop earrings, clip-on earrings and button earrings. Beaded, stone-studded, wired, pearled, traditional neck chains and necklace are designed in variety of styles and color themes. Bangle or bracelet designed in wood, ivory, silver, gold are the traditional fashion accessories matched with traditional ethnic attires. Bracelet styles comprise logo engraved or etched bracelets, bangle bracelets, cuff bracelets, beaded bracelets, leather bracelets, link bracelets, stretch or band bracelets, multi-chained bracelets, wristwatch bracelets and silver bracelets.

Figure 2.10 Embellished fashion jewelry

2.2.4.2 Hair accessories

Hair accessories are ornamental entities which can be attached to hair by different means like wrapping, tying, twisting or inserting. A consideration of religious significance, social class, age group, and level of fashion awareness is required while choosing materials for designing hair accessories. Hair accessories vary in their shapes, sizes, styles, materials and method of attachment to hair. Hair rings or bands, ribbons and bows, hairpins, hair combs, barrettes, beads are some of the commonly used hair accessories by females of all age groups. Hair rings and hair bands are cylindrically shaped hair accessories designed to confine hair strands and holding hair away from the face. Ribbons and bows are narrow fabric or elastic strips or braid wrapped and knotted around the hair. Hairpins are single-pointed pins which fasten the hair. Copper, wood, ivory and bones can be used in designing hair pins. Barrettes are long metal pins used to secure and fasten the hair with an underlying spring clip and a beaded head and guard cap. Threads or yarns assembled into an open, gauze like fabric creates netting which is generally used for binding long hair at the base of neck.

Hair forks, hair spikes, hair sticks and accessories comprising of unconventional materials like shells, coins, jewels, flowers, feathers, cow horns, bones and sheepskin are also preferred by females to add style statement to their tresses.

Figure 2.11 Hair accessories

2.2.5 High-tech accessories

The stature of fashion accessories is not merely restricted to enhacement of aesthetic appeal and complimenting the look of apparels. Nevertheless, the incorporation of technology in design and development of fashion accessories has revolutinsed the accessory industry taking a dig towards smart, wearble, technology driven fashion accessories that can be linked to smartphone, charged by solar power, built with touch sensors, or those enabling contactless payment. High-tech fashion accessories are becoming a promising and key growth opportunity as wearable technology is making entry in main stream prompting several retailers and brands to embrace this technology in designing of trendy and smart accessories. Smart jewelry, smart watches, handbags and shoes are some of the high-tech fashion accessories which may appear as ordinary piece of accessory at the onset but exhibit unimaginable and fascinating functionalities owing to technology that goes into development of these accessories. Smart jewelry serves as a means of combining fashion forward designs with the latest innovations in technology. Smart jewelry and watches are built and programmed to monitor stress levels, guide wearer in meditation, track sleep and activity status, play music, can easily track steps taken, calories burned, and distance traveled. The smart accessories so developed are particularly gaining huge applaud and acceptance in sports arena where monitoring body physiological parameters is of utmost importance and real time monitoring is plausible while the athlete or sportsperson is in dynamic conditions wearing a smart watch or a pair of smart sneakers.

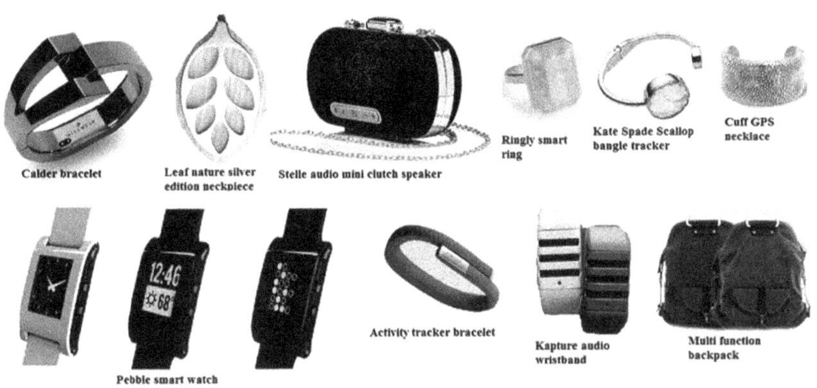

Figure 2.12 High-tech fashion accessories

Altruis by Vinaya is a smart jewelry collection comprising necklaces, rings and bracelets capable of connecting to smartphone and producing vibrating alerts that can be customized as per the wearers' needs. Calder bracelet is another piece of smart jewelry crafted out of Rose Gold, Gold and Palladium. It serves as an activity tracker on daily and weekly basis, keeping the wearer safe by instantly alerting emergency contacts with a three-tap system, and receiving notifications. Leaf nature silver edition neckpiece is built and programmed to track and monitor sleep, activity status and stress levels. Moreover, the smart jewelry can also keep a track of female's menstrual cycle. The salient feature of this smart neckpiece is that it does not require charging, since the replaceable battery lasts up to 6-months with an additional provision of backing-up the tracked and recorded data. Stelle audio mini-clutch is a smart handbag with a built in speaker phone and designed to play music for up to 15 hours. Ringly smart ring can track wearer's steps taken, calories consumed and distance travelled.

2.2.6 Popular accessory brands & unconventional accessory trends

Apparel brands in the pursuit to diversify their product range and offer one-stop-shop experience to their loyal customers are venturing in accessory segment and have been quite instrumental in exploring a range of innovative, technology-laced, aesthetically-appealing and functional accessories. The national brands operating at entry-level have been able to create niche for middle-income group consumers. The accessory brands which have successfully garnered the attention of masses are the ones who work by the notion of "Global in flavor and local in connect". The consumers can easily visit for window shopping or otherwise owing to the location of major brand outlets in catchment areas.

Almost all major apparel brands are expanding their product range offering an assortment of complimentary accessories considering the massive potential of accessory industry.

Table 2.2 Accessory brands

Category of fashion accessories	Popular brands
Handbags	Chanel, Gucci, Louis Vuitton, Prada,Dior, Kate Spade, Versace, Tommy Hilfiger, Calvin Klein, Hidesign, Lavie, Caprese, Baggit, House of Tara, Lino Perros, Peperone, Kara
Time wear	Rolex, Omega, Zenith, Titan, Citizen, Timex, Casio, Guess, Tommy Hilfiger, Fastrack, Fossil, Giordano, Daniel Klein, Rado, Emporio Armani, Chanel, Gucci, Armani
Men's footwear	Reebok, Bata, Nike, Adidas, Liberty, Woodland, Valentino, Converse, Lancer, Lotto, Red Tape, Puma, Paragon, Relaxo, Action, Louis Vuitton, Allen Edmonds, Aldo, Dolce & Gabbana
Women's footwear	Jimmy Choo, Walter Steiger, Gucci, Aldo, Dior, Prada, Burberry, Steve Madden, Kate Spade, DKNY, Tommy Hilfiger, Hush Puppies
Scarves	Gucci, Hemes, Chanel, Prada, Autumn Cashmere, Missoni, DKNY, H&M, Louis Vuitton
Eyewear	Ray-Ban, Tom Ford, Prada, Chanel, Dior, Dolce & Gabbana

The dominance of unbranded players in accessory industry further presents a scope for high-end brands to explore this category further. Accordingly, brands like Louis Vuitton, Hermes, Swarovski have been pioneer in launching their exclusive accessory line in India. Chanel, Gucci, Prada, Dior, Kate Spade, Versace, Tommy Hilfiger, Calvin Klein, Hidesign, Lavie, Caprese, Baggit, House of Tara, Lino Perros, Peperone, Kara are known for their exorbitant, elegant handbags while Jimmy Choo, Walter Steiger, Gucci, Aldo, Dior, Prada, Burberry, Steve Madden, Kate Spade, DKNY, Tommy Hilfiger, Hush Puppies are renowned brands popular among millennials for their exclusive range of footwear. Table 2.2 and Figure 2.13 highlight the major accessory brands.

Figure 2.13 Accessory brands

Unconventional accessory trends

The fashion industry is undoubtedly fad driven dynamic industry with apparel and accessory trends changing constantly every season and even between the seasons. The shapes, silhouttes and colors of accessories witness changes much like the apparel industry. Accessories no longer are categorized as mere add-ons rather they have become an indispensable part to complete the wardrobe look. The narrow approach of buying accessories matching the outfit is bygone in present times. The consumers' perception of accessories just serving a complimentary attribute to apparels is changing with more and more users embracing capsule collection mixing and matching accessories. Funky handbag or runway necklace style chic clutch bag paired with denims can add interest and punk to an old or boring outfit.

Furthermore, toaday's consumer is inclined towards sustainable, upcycled and recycled accessories for their apparel appreciation. Needless to mention that some accessories are designed with so much conviction considering the main stream consumer that consumers are fostered to buy the accessory first and later on buying coordinated clothing. The recent times have witnessed transformation from exorbitant accessories to affordable ones. Moreover, more and more users are switching to natural and organic material based accessories with jute, bamboo and straw handbags, mushroom and corn leather footwear gradually hitting the fashion world globally. The trends for accessories like apparels follow a cyclic pattern with the most endorsed and poularised ones embellishing the shelves and window display of retail stores while some others whose fate ends at fashion events and ramp walks.

It is thus imperative for accessories to be on-point along with apparels for rendering a unique style statement and an ever lasting impression. Chic belt bags, baker boy hats, red thigh-high boots, matrix glasses, shoulder grazing earrings and short strap cross body bags have been talk of the town and likeable assortments amongst the best street style looks of the Spring/Summer 2018 show season spanning from Milan, Paris to New York and London. Likewise, Spring Summer 2022 collection flaunts of timeless classics and bold fashion accessories such as net beret, wide waist belt, purple bags, green accessories, layered double chain necklaces, chucky platform high heel boots, pearl jewelry, gold hoop earrings, head scarves, etc. (Figure 2.13). The latest trends in fashion accessories are being discussed in the following section:

Baker boy hats – are must-have hats of the season witnessing an official come back with an improved look. The classic, casual cap has been spotted on chic ladies all over the world and has successfully replaced the wide-brim styles from the fashion arena.

Belt bags – are continuing to dominate the world of street style with their unique aesthetic and practical appeal. These hands-free handbags blend function and fashion and add a glamorous utilitarian touch to any outfit. Belt bags are available in a range of styles and designs ranging from sporty bum-bags to sleek, structured designs, and can be worn in several ways.

Red thigh-high boots – are taking centre stage and can complement well with denim shorts or a shift dress.

Matrix glasses – not only shield wearer's eyes from sun but also create a style statement. Retro sunglasses are slim, narrow and sharp designed with sleek black frames. The glasses have been an instant hit among celebrities and street style stars alike.

Shoulder-grazing statement earrings – Statement earrings have been the must-have jewelry item for the past couple of seasons and their craze among millennials suggests that they are going to rule fashion arena in the near future. The eye-catching earrings have been redesigned to reach upto shoulder-grazing lengths.

Short strap cross body bag – is a chic bag style featuring an adjustable or short strap and generally worn across body for the bag to sit at or just below the bust.

Figure 2.14 Trending fashion accessories in Spring/Summer 2022 collection

Figure 2.15 Baker boy hats

Figure 2.16 Belt bags

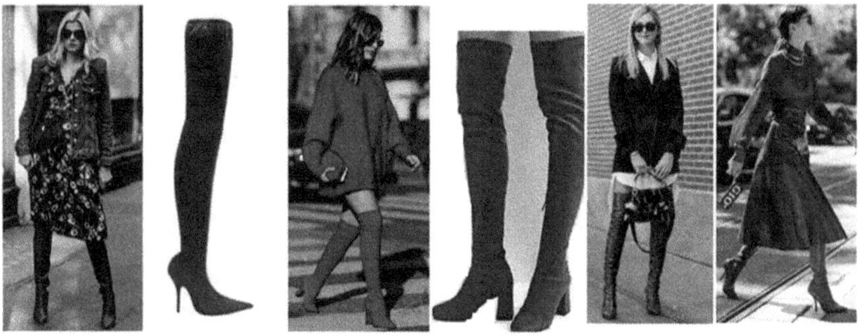

Figure 2.17 Red thigh-high shoes

Figure 2.18 Matrix glasses

Figure 2.19 Shoulder grazing earrings

Figure 2.20 Short strap cross body bag

Figure 2.21 Trending hair styling & accessories

3

Raw Materials for Fashion Accessories

3.1 Textile materials

A gamut of raw materials of both textile and non-textile origin are extensively utilized in accessory industry. The functionality, aesthetic appeal and varied styles of fashion accessories dictate the selection of raw materials for designing and construction of fashion accessories so as to achieve trade off between functional and decorative aspects alike. Raw materials also referred to as feedstock or unprocessed material, are the basic and starting materials for apparel and accessory manufacturing. Unprocessed or minimally processed materials serve as raw material for manufacturing finished goods. Raw material selection is a crucial factor that affects functionality, performance and aesthetics of developed fashion accessories. Raw materials of substandard quality will, in turn, affect the cost and the quality of end-products developed. Contrastingly, the cost of end-products developed from good quality, expensive raw materials will rise but it assures good quality and durability of manufactured accessories.

The assortment of fashion accessories such as footwear, handbags, gloves, knitted accessories and jewelry have been discussed in detail in Chapter 2. This chapter will give a detailed review of various textile and non-textile raw materials, generally used in accessory designing.

The inherent characteristics and properties of textile materials make them a preferred choice for accessories which are intended to be in close contact with skin and those that demand comfort, sweat absorption and thermal insulation, e.g. socks, gloves, mittens, etc. A gamut of textile fibres-natural and synthetic, yarns-carded, combed, ring, rotor, fancy yarns, chenille, boucle and fabrication techniques – weaving, knitting and non-wovens are experimented with and utilized with all success in designing of fashion accessories. The following sections will discuss the textile materials as raw materials for the accessory industry.

Non-textile materials like leather, fur, metals, precious and semi-precious stones also enjoy lion's share of accessory industry, with leather and fur primarily finding applications in footwear, handbags, gloves and belts categories, while latter are widely used in construction of watches and jewelry.

Although the accessory industry witnesses the dominance of textile and non-textile raw materials, however, the significance of different trims and notions in augmenting the look and functionality of accessories can not be undermined. Trims like decorative buttons, artificial flowers, appliques, sequins, Swarovski, laces, ribbons, tassels, etc. are extensively used for beautification and surface ornamentation of accessories. A scalled lace used for edge neatening of pair of gloves render them aesthetic appeal, likewise embroidered, painted appliques are finding their way on designer handbags and footwear. Furthermore, fashion jewelry designed from decorative buttons, tassels and other such trims are trending among youth and street fashion promoters. Apart from the decorative components, the functional entities - notions hold equal significance as far as designing and construction of fashion accessories are concerned. Zippers strategically placed in ankle and knee length footwear assist in easy opening and closing of the footwear. Likewise, velcros, buckles, zippers and eyelets attached on handbags ensure efficient closure to the bags.

Textile materials

Ranging from fibres of natural and synthetic origin to staple, filament and novelty yarns and woven, knitted, non-woven and braided fabrics are crucial raw materials, extensively utilized in designing and construction of fashion accessories.

3.1.1 Fibres

Fibre, a hairlike strand of plant or animal tissue, is the fundamental element of textile fabrics and other textile structures. Fibre can also be thought of as a unit of matter possessing length at least hundred times than its diameter. The fibres, in order to be used for fabric production and ultimately apparel and accessory industry, should qualify certain parameters which are endemic to textile structures. Textile fibres should be spinnable so that they can be converted into yarns of desirable count which, in turn, can be utilized for fabric production. The factors dictating the development and utilization of fibres include their spinnability, availability in sufficient quantities, economy in production and ability to impart desirable properties like warmth, softness and pliability to the end-users. The characteristics and properties exhibited by fibres depend largely on their source of origin. Fibres vary in their origin and can be natural or man-made. The natural fibres, as the name suggests, are the fibres which are derived from flora and fauna. Man-made fibres on the other hand are manufactured fibres like rayon, acetate, polyester, nylon, acrylic, metallic, olefin rubber, modacrylic, etc. The detailed classification of fibres based on their source of origin will be discussed in detail further in the chapter.

The fibres irrespective of their origin should possess certain properties to be useful for textile applications. Natural as well as synthetic fibres should have the length several hundred times more than their width, must be spinnable to be readily converted into yarn, and should be strong enough to withstand any mechanical action to which they are subjected during production. Textile fibre must be at least 5 mm of length in order to be supple, flexible and strong, to be efficiently spun into yarn. The fibre properties like elasticity, fineness, uniformity, durability, luster and crimp are ofcourse crucial but would also depend on the intended application area of the fibre.

3.1.1.1 Fibre properties

The properties which qualify the textile fibres to be used for textile applications are being discussed in detail in the following section:

(a) *Length and appearance*: length of fibre influences their selection for fabric production. Short staple natural fibres like cotton, jute, wool impart rough look to fabrics made from these fibres. However, fabrics made of nylon, polyester filament yarns exhibit smooth and shiny look. Moreover, short wavy fibres like cotton and wool have affinity for dirt particles while long, continuous filament fibres like silk or synthetic fibres are easy to clean and do not get as dirty as their natural, staple counterparts. Accordingly, wool and jute fibres are used for fashion accessories where designer aims to encash rough, crimpy fabric appearance as the design feature while accessories like scarves and stoles which demand sheen and smooth, soft look and feel, utilize silk or filament fibre yarns and fabrics in their designing.

***(b) Moisture absorption*:** the fibres vary in their affinity for water molecules with natural fibres like cotton inherently hydrophilic while synthetic fibres like polyester, nylon exhibit hydrophobicity or low moisture absorption. The moisture absorption property of fibres determine their selection for apparels and accessories suitable for a particular season. Summer wear apparels and accessories should provide rapid sweat absorption from wearer's skin and thus utilizes hydrophilic fibres like cotton. Fabrics made of synthetic fibres, being hydrophobic do not soak sweat and are uncomfortable in summers. A special consideration of fibre type is important while designing accessories like gloves, socks which are in intimate contact with skin where sweat accumulation may lead to dampness and discomfort to the wearer.

***(c) Thermal conductivity*:** the ability of the fibre to conduct heat away from the wearer's skin is yet another crucial factor influencing the selection of fibre for a particular end-use. Cotton and rayon exhibit high thermal conductivity and have the ability to conduct heat from body, lower the body temperature and providing cool feel next to skin, Accordingly, fibres with higher thermal conductivity are preferred for accessories intended for summers like driving gloves, socks, stockings. However, the requirements in winter are quite contrasting with such material selection that can provide thermal insulation and prevent the body heat to escape out. Therefore, fabrics made of wool, acrylic and synthetic fibres, being poor heat conductors of heat aid in keeping the wearer warm in winters.

***(d) Strength*:** The strength of fibres determine the ease of washing of fabric made from the chosen fibre. Fibres have property of strength variation in dry and wet states. Fibres like wool, silk, rayon loose strength in wet state while cotton and synthetics exhibit high strength even when wet enabling easy laundering of fabrics made from such fibres. Therefore, cotton and other synthetic fibres should be preferred for daily wear apparels and accessories which require frequent cleaning and laundering. However, occasional wear accessories like silk scarves, stoles are made of delicate fabrics prone to strength loss on repeated washings and need dry cleaning and less frequent washing cycles.

3.1.1.2 Classification of textile fibres

The various types of fibres used in the textile industry exhibit their own individual, unique properties. The properties fibres display, depend largely on their source of origin. The fibre properties desirous for textile applications have been discussed so far. This section is devoted to elaborate classification of fibres based on their nature and origin, botanical, zoological or chemical name, moisture absorption properties, thermo plastic behavior and utility.

(a) Classification based on origin

The primary classification of fibres is based on their origin and thus can be classified as natural and man-made fibres.

Natural fibres

Natural fibres, as the name suggests are obtained from nature. The natural sources for these fibres can be plants, animals or minerals. Figure shows the classification of vegetable based cellulosic fibres. The fibres obtained from plants include cotton, kapok (from seeds), sisal, banana, pineapple (obtained from leaves) and bast fibres like jute, flax, hemp, kenaf, ramie, etc. (obtained from plant's stems). Cellulose fibre extensively used in accessory industry include cotton, linen, flax, hemp and jute. A variety of fashion accessories ranging from cotton canvas satchels, tote bags, tapestry luggage to sports socks, sneakers, bandanas, handkerchiefs, scarfs, stoles, hats and caps uses cotton as the major raw material.

The salient features of cotton fibre that makes it suitable for a range of apparels and accessories include its softness, breathability, moisture absorbency and temperature regulation property. The natural textured surface of fibre in addition to its moisture absorption provides comfortable feel and breathability next to skin. Accordingly, the fibre is preferred for applications where rapid water absorption is the prime concern like terry-cloth apparel such as beach coats. Moreover cotton is versatile, easy to care and handle fibre that is not just static free, hypoallergenic, and pill free but also has the ability to retain its original feel and color. The above stated attributes make cotton year round fibre suitable in both warm as well as cold weather.

Another natural fibre that is gaining interest among accessory designers for designing footwear and handbags is linen. Hemp and jute are rope like fibres and exhibit coarseness compared to linen and generally preferred for accessories that demand rough texture and durability. Hemp, jute are finding applications in accessories like fashion jewelry, handbags, belts, hair accessories, footwear, bag-pack, tote, gunny-bags and mobile covers.

Jute is a plant based multicellular fibre characterized by nodes and cross markings in longitudinal view and polygon shapes in cross-section. Jute is hygroscopic in nature with moisture regain of 12-14% and its color varies from yellow to brown to grey and is good insulator of heat and electricity. Jute fibre is used for making burlap, hessian, gunny cloth which serve as raw materials for accessories like handbags, jewelry, etc.

Bamboo and straw are another plant based fibres gaining popularity among accessory designers. Bamboo is wood like tropical grass while the sources for straw are wheat stalks, grasses, sisal hemp and rice paper. Summer wear hats, hair accessories, different handbag styles like totes, slingbag, clutch are designed from straw, bamboo or strips of bamboo woven like straw. The pliability and sturdy hand of bamboo canes make them suitable for gripping elements and handles for handbags. However, straw is generally preferred for delicate fad accessories trending in a single season which is accounted to its inferior durability and flex abrasion resistance. Repeated bending in one position may cause fibre breakage hampering the aesthetic appeal of accessory and rendering it useless.

The fibres such as wool (obtained from sheep) and silk (obtained from silk worm cocoons) are protein based fibres and finding applications in a variety of accessories. Pashmina (cashmere or cashmere/silk blend), wool and silk are used for designing exclusive and high-priced stoles, scarves and other winter wear accessories. Pashmina is obtained from the underbelly of the Capras goat found in India's Himalayan mountains. Wool is easily distinguishable from hair or fur fibres showcasing crimpy appearance and elasticity. Fabrics made from wool fibre exhibit durability owing to tear and snag resistance and antipill properties. Moreover, wool fabrics provide easy care and handling properties, drapes stunningly, and maintains resilience in wet condition. The most crucial property of wool fibre of providing insulation by holding air layer next to the skin makes the fibre a preferred choice for winter wear apparels and accessories. Henceforth, winter wardrobe is incomplete without the inclusion of wool-based hats, scarfs and mittens.

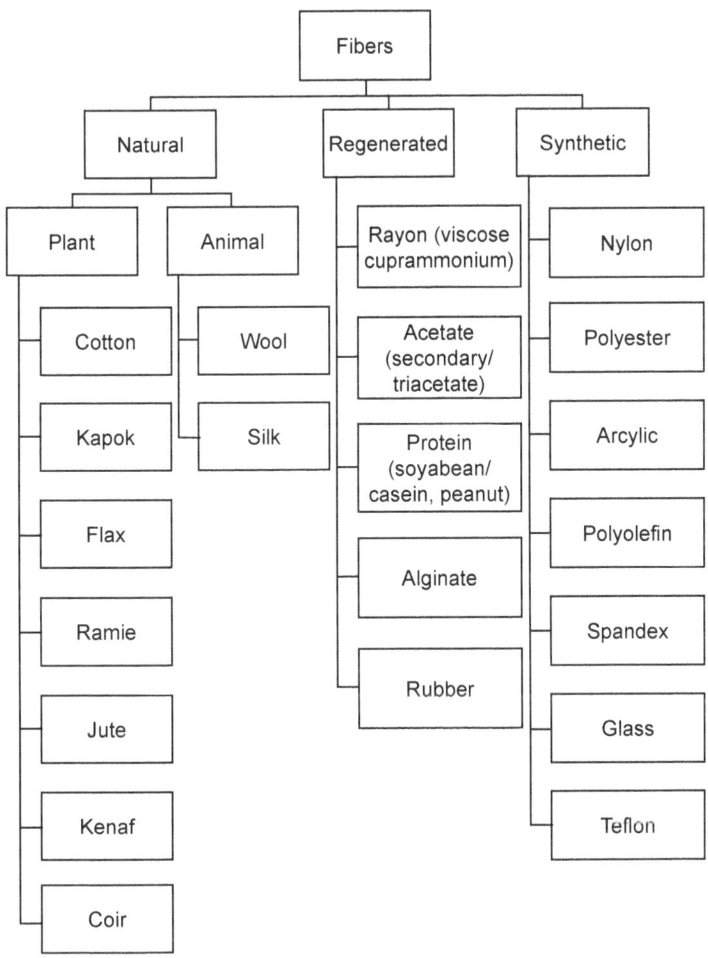

Figure 3.1 Classification of fibres

Another natural protein fibre namely silk is composed of fibroin and is the product of insect larvae that forms cocoon. The most exclusive variety of silk is obtained from the cocoons of the larvae of the mulberry silkworm Bombyx mori. Silk is considered to be the most luxurious of all fibres characterized by unique natural luster. Triangular prism like structure of the silk fibre is responsible for its shimmering appearance allowing silk fabric to refract incident light at different angles, thereby producing varying color effects. The silk fibre has an additional advantage of being one of the strongest natural fibres and offers remarkable abrasion resistance with many years of service to wearer.

Furthermore, the fibre provides extremely soft texture, is elastic and displays the ability to retain its original shape with minimal shrinkage.

The unique properties of this protein based fibre available in filament form make it the most anticipated option for apparel and accessory designers to render a loyal, pleasing look to end products. Silk fibres used in accessories like stoles, handkerchief, pocket square, bow, scarves, men's ties, brimmed hats, stocking and gloves offer smooth, lustrous and pleasing appeal. The fabrics produced from this luxurious fibre however, need special care as far as dry cleaning, washing and pressing is concerned. Manual washing of silk apparels and accessories is usually recommended unless the end products have been processed for machine laundering. Moreover, the exorbitant prices of silk fibre prompt designers to look for cheaper substitutes that resemble silk, such as polyester and nylon microfibres.

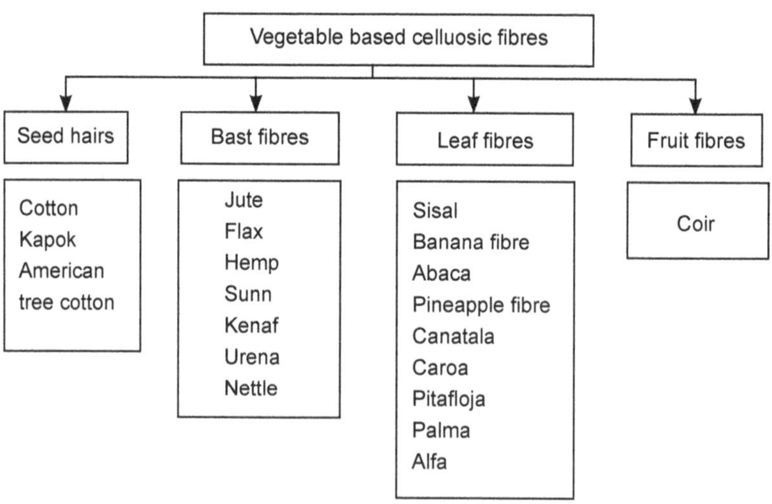

Figure 3.2 Classification of vegetable based cellulosic fibres

Figure 3.3 Natural fibres

Accessory designers like to experiment with an assortment of fibres available in different price ranges to exploit their naturally inherent textures, soft feel and unique appearance. Figure shows the classification of various hair fibres like wool, fur, camel and goat hair. The expensive hair fibres like cashmere and vicuna are designers' first choice while designing winter wear accessories like stoles, scarves, gloves and varied knitted accessories for elite social segment. Vicuna, like cashmere is an expensive fibre whose thin, soft hair comes from the vicuna of South America. Angora rabbit hair being extremely fine is also being used for winter wear accessories as a cheap substitute for cashmere The only drawback associated with the fibre is its shedding tendency. Angora goat is the source of another hair fibre, mohair which is known for its strength, resilience and lustre. Alpaca and camel's hair are other hair fibres which are widely used for coats, suits, sports coat and accessories intended for winters. Alpaca presents silky appearance and is known for hair length longer than sheep's wool while camel's hair provides wool like texture to apparels and accessories designed from this fibre.

Vegetable fibres like ramie, jute and hemp in contrast to hair fibres are economical and are characterized by rigid feel, coarseness and brittleness and thus find application in designing accessories where strength, abrasion resistance and rough texture are the requisites. The fibres are thus generally preferred for designing lower priced accessories like fashion jewelry, belts, handbags. Asbestos is the natural mineral fibre known for its fire-proofing and insulating properties. Accordingly, the fibre was utilized for flame retardant protective clothing. However, the fibre has limited application in textile arena in recent times owing to its carcinogenic nature and hazards to lungs by continued inhalation of asbestos fibre.

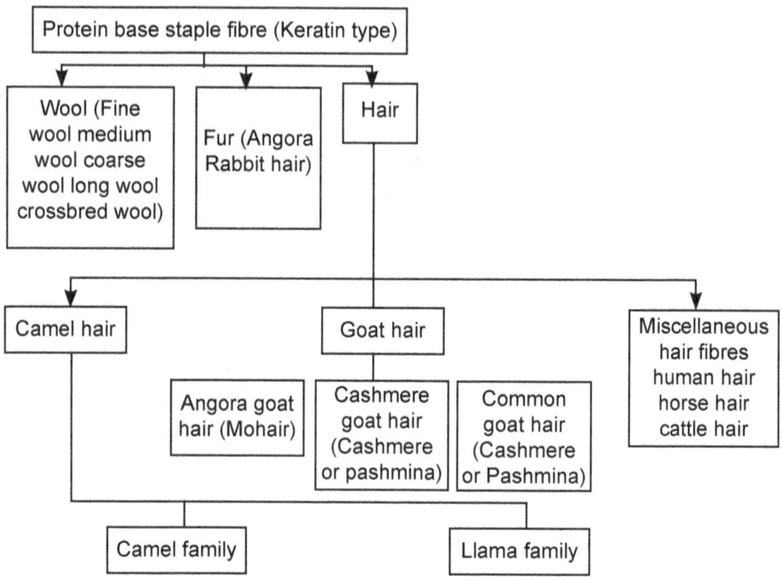

Figure 3.4 Classification of Protein base staple fibres

Classification of hair fibres

Man-made fibres

Man-made fibres also referred to as synthetic or artificial fibres are the fibres that are developed by mankind to meet the ever-increasing demands of fibres in apparel and accessory industries since the sources of natural fibres are on the verge of depletion. Moreover, the stringent animal rights discourage slaughtering of animals for their skin, fur and hair fibres. Therefore, the man-made fibre industry is bound to grow tremendously. The man-made fibres are manufactured with the aim of cutting down costs associated with natural fibres and at the same time achieving desirable properties like high strength, abrasion resistance, soft feel, drapeability and varied textures. Generally, fabrics produced from man-made fibres are named similarly as those produced with natural fibres such as batiste, is synonymous with cotton, flannel with wool; satin and shantung with silk fabrics and crash is heavy weight linen.

Manmade fibres are classified into regenerated, synthetic and miscellaneous inorganic fibres on the basis of raw materials employed for their manufacture. Regenerated fibres like viscose rayon, cuprammonium rayon, acetate, triacetate, casein, ardil, rubber (Figure 3.5) are the man-made fibres that belong to cellulosic group and are produced using natural polymers, i.e., cellulosic base materials thus requiring minimum chemical steps.

Acetate it has trade name of Ariloft, Chromspun had a luxurious feel and appearance, is available in a wide color range, drapes extremely well, is relatively fast drying and is resistant to shrinkage, mildew and moth. It is used for blouses, dresses, foundation garments, lingerie, linings, shirts, slacks and sportswear. Fabrics knit with the acetate fibre include brocade, crepe, double knit, knit jersey, lace, taffeta and tricot.

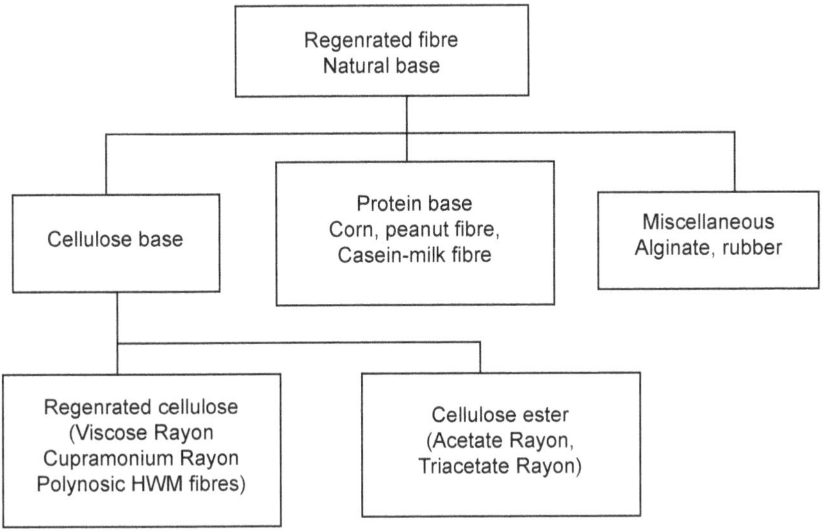

Figure 3.5 Classification of regenerated fibres

The non-cellulosic group include fibres like nylon, polyester, acrylic, polyolefin, spandex, glass and teflon which are protein based fibres (Figure 3.4). These fibres constitute of molecules of carbon, hydrogen, nitrogen and oxygen derived from petroleum and natural gas. The fibres manufactured by synthesizing various chemicals like the petroleum products are classified as synthetic manmade fibres. Natural raw materials are not required as a base material for their manufacture unlike the regenerated ones.

Nylon fibre belonging to this group is manufactured using hexamethylenediamine and adipic acid while polyester uses dimethyl terephthalate and ethylene glycol in its production. The process basically involves conversion of chemicals into fibre forming substances which can be drawn into filaments of required coarseness or fineness.

Dacron, Fortrel, Trevira and Ultra touch are the trade names of one the mostly widely used synthetic fibre - polyester in apparel and accessory industry. The popularity this fibre enjoys in textile arena is attributed to its high strength, resistance to stretching, quick drying, retention of heat set pleats and creases and ease of laundering. Accordingly, the fibre has been extensively used in variety of apparels and accessories namely blouses, shirts, suits, children's wear, hosiery, insulated garments, ties, scarves, stole and casual handbags. The new breed of polyester produced under microfibre group with its salient features of being inexpensive, drapeable, wrinkle free, easy handling and laundering is presenting tough competition to silk fibre fabrics which are comparatively expensive and requires much care as far as laundering and dry cleaning is concerned. The additional advantage of micro polyester is its color fastness and ability to take up prints with greater clarity, a much sought after property for designing printed apparels and accessories.

Nylon is known for its exceptional strength, abrasion resistance and easy care properties making it desirable fabric for hosiery, blouses, dresses, lingerie, underwear, raincoats, ski wear and suits.

Acrylic is also referred to as Acrilan, Orlon and Zefran. The fibre is characterized by its soft feel, warmness, wool like texture and light weight. Moreover, the acrylic fibre fabrics display considerable strength, resilience, drying ability, and are resistant to moths, sunlight, oil, and chemicals. Therefore, fabrics made of acrylic fibre namely fleece, fake furs, jerseys find application in winter wear and knitted accessories like mufflers, cowls, stoles, beanies, socks and gloves along with apparels like dresses, infant wear, knitted garments, skirts and sweaters. Modacrylic, another synthetic fibre is used for fabrication of fleece and knit pile fabrics. The fabrics owing to their softness, resilience, abrasion and flame resistance, drying ability, and shape retention, are preferred for deep-pile coats, linings, simulated fur accessories and hair accessories.

Olefin is another organic man-made fibre that exhibits excellent colorfastness, extreme strength, resistance to mildew and perspiration and is thus, is garnering designer's attention for designing of knitted accessories, winter wear and sportswear apparels and accessories.

Another fibre that has taken the apparel and accessory industry by storm is rayon fibre which offers multitude advantages of being soft, comfortable and easily dyeable. Consequently, the fibre is preferred choice for a variety of apparels and accessories like sportswear, blouses, jackets, lingerie, rainwear, shirts, scarves, stoles and ties.

Spandex Produced extensively by Dupont as Lycra, its major advantage is its ability to be stretched 500 percent over and over without breaking, always returning to its original length. It is lightweight, more durable than rubber, and resistant to body oils, it is used wherever stretch is required such as in athletic apparel, bathing suits, foundation garments, ski pants and sportswear.

Miscellaneous inorganic group comprises of fibres which use substances like metal and glass in their manufacture. The malleability and ductility of the inorganic fibres render them usable for textile applications. However, high cost and technical difficulties offers hinderance to wide spread acceptance of the fibres.

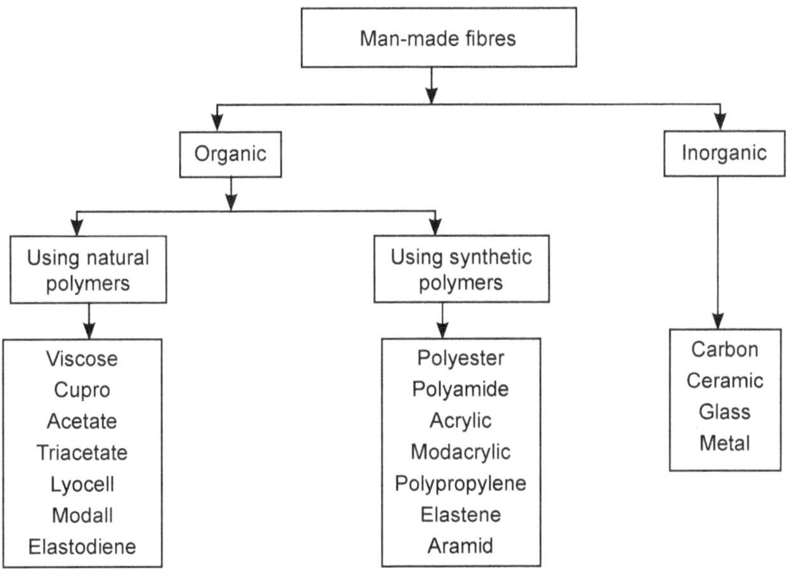

Figure 3.6 Classification of Man-made fibres

High-performance & speciality fibres

High-performance fibres are quite distinct from staple fibres in possessing exceptionally high tenacity, stiffness and resistance to bending and stretching. The unique properties of these fibres make them suitable for high-tech, smart apparels and accessories like protective clothing, fire-fighting suits, sportswear, etc. Polyphenylene Benzobisoxazole (PBO), Polybenzimidazole (PBI) Melamine Novoloid, para-aramids and gel-spun polyethylene belong to the category of high performance fibres.

Polyphenylene Benzobisoxazole, PBO fibre is composed of polyphenylene benzobisoxazole polymer. PBO is a rigid rod polymer exhibiting highly ordered crystalline and rigid structures which accounts for the fibre's high strength, modulus and excellent thermal stability. Heat resistance and light weight of PBO makes it a favorable choice for protective clothing and accessories like head gears, covers, gloves, etc. for fire-fighters.

Polybenzimidazole (PBI) is composed of long-chain aromatic polymer having reoccurring imidazole groups in its polymer chain. The fibre offers low modulus and high moisture regain, thus imparting comfort to apparels and accessories designed thereof. The desirable properties of PBI are being exploited in designing of military clothing and protective gloves replacing the earlier used asbestos fibre which is responsible for health hazards to wearer.

Melamine is a thermosetting manufactured fibre composed of atleast 50% by weight of a cross-linked melamine polymer. The rigid structure of the fibre is responsible for its low thermal conductivity and high heat resistance with an additional advantage of being resistant to chemical degradation and suitable for fire-fighting suits and protective accessories.

Novoloid is yet another high-performance manufactured fibre constituted by atleast 85% by weight of a cross linked novolac which is a three dimensional amorphous network of phenol-aldehyde. The fibre is suitable for heat resistant clothing and accessories owing to its low thermal conductivity.

Vinyon is a is composed of long-chain synthetic polymer with atleast 85% by weight of vinyl chloride units. The fibre is extensively used in underwear, socks, flame resistant draperies and as heat-sensitive binder in non-woven applications.

Classification according to their botanical, zoological or chemical name

Another classification of fibres is based on their botanical, zoological and chemical names. Botanical classification includes vegetable fibres like cotton, jute, flax, etc. while zoological classification includes animal fibres like wool, silk and hair fibres. Manmade fibres are classified based on their chemical names.

Classification according to moisture absorption property of fibres

The moisture absorption property of fibres determine fibre's ability to undergo different wet processing treatments, i.e., bleaching, mercerizing, dyeing and finish application. The fibres exhibiting affinity towards moisture are referred to as hygroscopic or hydrophilic fibres. hydrophilic groups present in hydrophilic fibres like natural fibres attract water molecules. However, hydrophobic fibres repel water molecules due to presence of very few water attaching groups in fibre's polymeric chain. The low moisture absorption characteristics of these fibres is attributed to absence of water attracting groups, in turn creating difficulty in dyeing and bleaching. The mechanical processing of fibres is troublesome owing to their hydrophobicity and tendency to develop static electricity unlike hydrophilic fibres.

Classification according to fibres' thermoplasticity

The textile fibres can also be classified into thermoplastic and thermosetting fibres. Thermoplastic fibres undergo deformation by application of heat and pressure, without any chemical change. The thermoplastic fibres can thus be softened, moulded and heat set on exposure to heat. Synthetic fibres exhibit thermoplastic properties. The fibres which do not possess the above characteristic are designated as non-thermoplastic.

Classification according to their utility

The fibres can further be classified into major and minor fibres based on their utility or application area. Major textile fibres are widely used in apparel and accessory industry like cotton, flax, wool, silk, jute, viscose rayon, acetate rayon, nylon, polyester, etc.

Minor textile fibres like hemp, ramie, jute, banana, abaca, asbestos, bamboo, soybean, pineapple, metallic fibres, milk fibres, casein, alginate, rubber find application in textiles to limited extent. The global concern for sustainability and quest for new innovations call for the experimentation with unconventional, minor fibres for varied textile applications.

3.1.2 Yarns classification

Yarn can be described as continuous strand of natural, manmade spun or filament fibres that have been subjected to twisting process to provide strength and structural integrity. The yarn type and structure has direct influence on the resultant fabric's texture, appearance, durability and drape.

The yarns produced by various spinning methods viz ring, rotor, air-jet, friction spinning, etc. are thereby used for fabric production by weaving, knitting, crocheting, braiding, etc. Apparel and accessory industry uses a variety of yarns ranging from spun yarns to filament yarns, textured yarns, covered/wrapped yarns and fancy yarns like chennile, boucle and slub yarns (Figure 3.7). Yarns are generally categorized based on their physical structure and performance properties.

3.1.2.1 Classification based on structure

The primary classification of yarns based on their structure is - filament and short staple yarns. Filament yarns are composed of long, continuous strands of fibres. Monofilament yarns consist of single filament in yarn cross-section and present harsh and rough handling, thereby limiting their application in sheer fabrics, open work decorative fabrics, fabric webbing, etc.

Multifilament yarn as the name suggest consists of multiple filaments in the yarn cross section. The cohesiveness to multifilament yarns is imparted by twisting them together at varying twist levels. The amount of twist along with fibre characteristics like lustre, hand and cross sectional shape are some of the factors that govern yarn appearance and its feel. Texturing can be used to change surface texture of yarns and in turn the feel and bulk of filament yarns. Textured yarn is processed to introduce fine distortions such as loops, crimps or coils along the filament's length and is widely used in apparels and accessories where shape retention, insulation and bulkiness are the desirable properties. Consequently the yarn finds utility in knitwears, suits and overcoats, belts, bags, furnishings and upholstery.

Staple or spun yarns consist of short-length fibres and thus require twisting to hold them together for formation of long, continuous yarn. Short staple fibres like cotton, flax and wool have staple sizes ranging from ½ to 36 inches in length and require spinning for yarn formation to ensure fibre binding in the yarn structure. Man-made fibres generally produced as continuous filaments can be used for fabric production without undergoing the spinning process. However, when man-made filaments are used to substitute the natural spun varieties, the filaments are cut into smaller pieces or staples and then spun in a similar manner to natural fibres. Rayon produced artificially as filament undergoes spinning when it is intended to imitate linen.

The most common of all the available spinning methods is ofcourse ring spinning. Open-end spinning also referred to as rotor spinning is a more recent and faster spinning system compared to former.

Apart from ring spun carded yarns, combed cotton yarn is also frequently used for accessories where silkier, smoother appearance and better drape properties are desirous as combing process imparts the above stated properties to the combed yarn.

Cotton fibre obtained from plants is spun into yarn with S-twist arrangement. The range of cotton yarn count generally used for textile applications is Nes20 to Nes100. Blended yarns comprising of fibres of distinct properties are also being used for varied textile applications. Blended yarn is formed by the composition of two or more materials (like cotton, viscose, acrylic, wool, bamboo, Lycra, modal, and linen) in order to exploit the relative merits of each individual component. The count ranges for blended yarns vary from Nes10 to Nes 60. Mercerized yarns in count range of Ne20 to Ne120 are generally preferred for knitted and crocheted apparels and accessories owing to enhanced strength, dye affinity and luster properties of the mercerized yarn. Synthetic yarns consisting of man-made fibres like polyester, elastane, and acrylics are available in both staple and filament forms and widely used in winter wear and sports accessories apart from clothing and home furnishings. Acrylic yarn, another synthetic fibre yarn is produced with S twist pattern and 2 plied arrangement in count ranges from Ne10 to Ne50. Acrylic yarn is extensively utilized for winter wear apparels and accessories like sweaters, pajamas, innerwear gloves, scarves and stoles. Nylon yarn finds application for commercial purposes owing to its tenacity and resilience. Nylon yarn is generally produced in count range from Nes6 to Nes66. The yarn is versatile with varied application areas ranging from tire cords, ropes for nets to parachutes, belts, ropes and apparels, accessories and home furnishings. Viscose yarn, a cellulose based fibre yarn, is used as embroidery thread and novelty yarns for fashion apparels and women's lingerie owing to its high tenacity and colour fastness. Spandex yarn also referred to as lycra is known for its elastic nature and accordingly preferred for apparels and accessories that demand high stretchability like socks, innerwear, sweaters, swim suits, ski pants, golf jackets, gloves, elastic waistbands and wristbands, stretchable jewelry, pantyhoses, etc.

Apart from short staple natural and spun, filament synthetic yarns, speciality/ fancy yarns are also used in apparel and accessory designing in order to obtain a gamut of functional and aesthetic aspects. Apparel and accessory designers usually experiment with a variety of yarns with distinct surface characteristics to add visual delight to end products designed thereof.

Yarns with distinct surface texture, irregular profile or a construction unlike basic single and folded yarns are usually referred to as fancy, novelty or complex yarns. Complex yarns are characterized by more than one component unlike novelty yarns which can be single, ply or cord, staple or filament. The novelty yarns are weaker than other yarn types and are generally used in woven fabrics' crosswise direction where strength is not essential. However, the most common use of fancy yarns is to impart distinguished texture and decorative effects to knitted or woven fabric constructions. Spiral, gimp, slub, knob, cover, chenille and snarl yarns are some of the basic profiles of the fancy yarns. Figure shows the different fancy yarns generally used for textile applications.

Spiral yarn is formed by plying together of two single yarns of different thickness and twist level with slight over feed of course yarn.

Gimp yarn is produced by two techniques namely plying and the hollow spindle process. Plying involves twisting the component yarns in direction opposite to twist in either of the parent yarns while the latter uses two ground yarns on which the drafted ribbon is buckled in a wavy configuration. Snarl yarns also uses plying process using highly twisted yarns as the components. Chenille yarn is made by leno weaving in which two weft yarns are placed between each crossing of the weft yarns. Pile effect is produced by cutting weft length extended between the warp yarns. Chenille yarns are made by a totally different process and require several steps in their preparation. First, leno-weave fabric is woven. This fabric is cut into strips, and these strips, which have a soft pile on all sides, are used as yarns.

Boucle yarns or loop yarns are formed of multiple plies. An effect yarn creates decorative effect by forming irregular loops around base yarn. The effect yarn is bound or tied to base by another yarn. Boucle yarn is generally used for designing winter wear accessories like scarves, hats, and mittens. Flake or flock yarn is composed of loosely twisted yarns with fibres of varying colors in the structure and is used to impart tweed effect to fabric so produced. Nub yarns, also known as knot or spot yarns, are ply yarns characterized by the presence of an enlarged bump or nub in the yarn structure and is formed by twisting effect yarn around a base yarn multiple times in a confined area. Nubs are held in position by binder yarn and may be spaced at regular or irregular intervals along the yarn length.

Slub yarns are formed of staple fibres. Varying levels of twist in the yarn creating areas of looser twist imparts slub effect to the yarn. An irregular diameter along the yarn results in the formation of a long, thick, soft area in the fabric called slub. The slub yarns when used to weave fabrics will provide irregularities on fabric surface.

Winter scarves and mittens in chenille (fuzzy caterpillar) fabrics are examples of complex yarns in fashion accessories. These complex yarns are characterized by loops projecting from the axis of the yarns.

Core-spun yarns also referred to as core-sheath construction do not belong to category of novelty yarns, as the former consists of central core and wrapped or twisted exterior layer , both of which contain different fibres. Elastomer core like spandex wrapped on the exterior by another fibre can be used to produce core-spun yarn. Sewing threads are produced in core-spun construction with polyester fibre in core that contributes to yarn strength and cotton wrapping it as an outer layer providing high temperature resistance.

3.1.2.2 Classification based on number of components

Another classification of yarns can be based on number of components. Accordingly, the yarns have been classified as single, plied and corded yarns. A single yarn is made from a group of filament or staple fibres together. If a single yarn is untwisted, it will separate into fibres. A single yarn might be identified as either a single yarn of staple fibres or a single yarn of filament fibres. Plied yarns are made by twisting together two or more single yarns. Cord yarns are made by twisting together two or more ply yarns. Corded yarns are used in making sewing thread and are woven as decorative yarns into some heavy weight novelty fabrics.

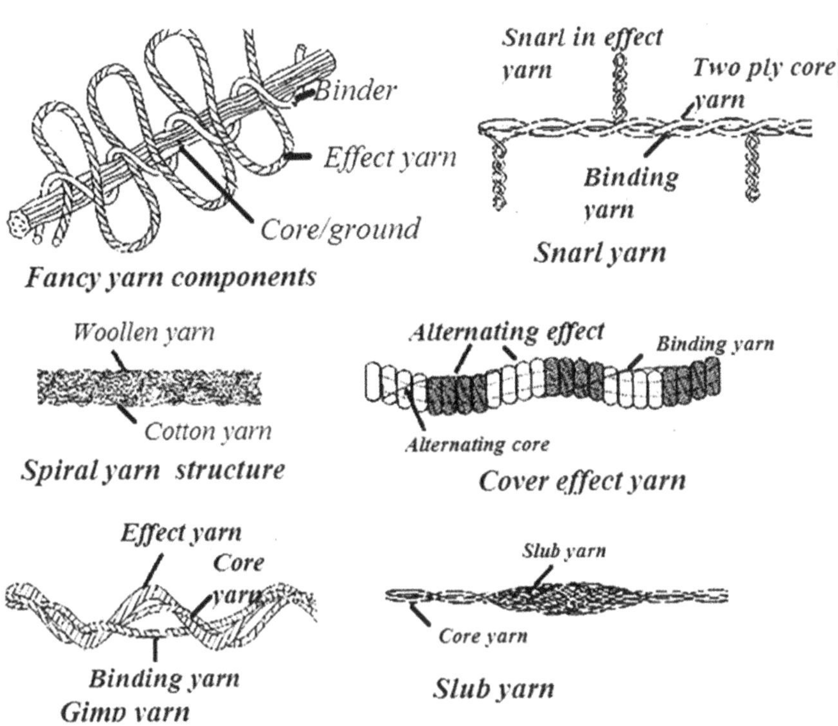

Figure 3.7 Different types of fancy/speciality yarns

The metallic yarns are used to provide extra sheen to apparels and accessories. The metallic fibres are made of metal, metal-coated plastic or plastic-coated metal. The metallic fibres suitable for textile applications should preferably be composed of delicate metal threads that can be spun, woven and knitted as per requirement. Metallic fibres comprise of gold and silver, aluminized nylon and aluminized plastic yarns. Other sources of metallic fibres and yarns are wires, bundles drawn from larger diameter wire, molten metal cast, etc. The basic manufacturing processes of metallic fibre include lamination and metalizing process. Fabrics utilizing metallic yarns are generally used for designing of apparels and accessories that demand extra sheer, particularly occasional, party wear attires and accessories like handbags, clutch, jewelry, etc. The metallic yarns can also be blended with other natural and synthetic yarns like wool, cotton, nylon, etc. The applications are innumerous with metallic yarns particularly those composed of stainless steel metallic fibres being used for e-textiles, protective clothing, cut resistant gloves, space suits owing to the conductivity, durability and strength offered by these yarns.

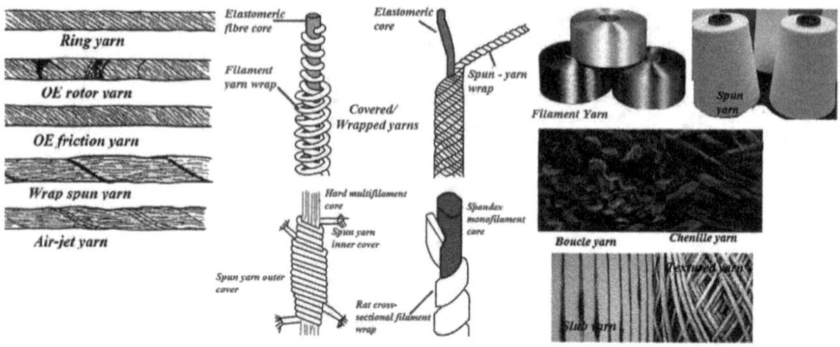

Figure 3.8 Different Yarn Structures

3.1.3 Fabrics

The yarn formation process is followed by fabric production. The process of conversion of yarn into fabric is referred to as fabrication. Durability, price, appearance and application area dictates the selection of fabrication method chosen for fabric production. Several fabrication techniques namely weaving, knitting and non-woven techniques are employed to produce fabrics in order to fulfill the demands of apparel and accessory industry. Felting, crocheting and bonding are yet another techniques frequently used for apparel and accessory designing. Knitted and crocheted fabrics are classified as looped fabrics which are formed by a series of interconnected loops. Knotted fabrics like lace, nets, macrame are a result of sequential or random knotting of yarns as per the design requirement and are trending in several accessories ranging from wallets, handbags to hair accessories, jewelry, scarves and stoles.

Each fabric type has unique, salient feature which makes it suitable for a specific application e.g. woven fabrics are characterized by high strength and dimensional stability and accordingly, extensively utilized for designing accessories that need to be sturdy, durable like luggage bags, bagpacks and footwear, etc. Footwear, handbags and belt manufacturers are substituting expensive leather with canvas, denim and other similar higher gsm fabrics that offer best of both worlds with strength, durability and affordability to consumers unlike leather.

Knitted fabrics are known for their stretchability, moisture transmission and comfort aspects. Additionally, the porous structure of knitted fabrics enable them to entrap still air in structure enhancing their thermal insulation. Hence, accessories worn in intimate skin contact like socks, stockings, gloves, mittens and winter wear accessories explore these salient features of knitted fabrics.

3.1.3.1 Woven fabrics

Woven fabrics are constructed by the interlacement of warp (lengthwise) and weft yarns (cross wise) at right angles. Weaving techniques are used to produce a variety of fabrics like plain, twill, satin, pile, jacquard and dobby. Woven fabrics offer durability (longevity) and dimensional stability to end products designed from them. Both the properties are very crucial for accessories that are subjected to frequent washing cycles and those that support excessive weight like handbags, bagpacks and luggage. The strength and durability of woven fabrics is fully exploited for designing smaller accessories like silk scarves, ties, belts, watchbands and ribbons.

Plain weave is formed by regular interlacement of the weft (filling) yarns over and under warp yarns. Plain fabrics of varying thickness can be manufactured to meet the requirements of apparel and accessory industry. Basket weave is the variation of plain weave and involves the interlacement of two or more sets of yarns as a single entity. The fabrics woven with basket weaves, however, pose a problem of shifting of yarns from their designated positions during apparel production or laundering. Batiste, lawn, voile, chiffon, buckram, organdy, crepe, grosgrain, cambric, poplin, flannel and taffeta are some fabrics woven in plain construction.

Twill woven fabrics are known for their durability and identified by diagonal line appearance in the fabric. The fabric is formed when each filling yarn interlace an over-one, under-two courses through the warp yarns. Denim, gabardine, glen check, foulard and whipcord are examples of twill woven fabrics.

Denim, a prime example of twill woven fabric is vastly used for designing accessories which are intended for rough and tough usage, handling and demand a certain level of strength and durability. Consequently, denim serves as a crucial raw material for handbags, wallets, footwear, etc.

Satin woven fabrics are characterized by smooth and shiny surface. Satin drill, venetian, satin-back gabardine, doeskin, duchesse satin-back crepe fabrics are some of the fabrics woven with satin construction. This woven structure varies from other woven structures as far as yarn interlacement patterns are concerned. The yarns interlacements does not occur at close intervals unlike in other techniques, however, interlacement occurs after a floating of yarn four to twelve times. The floating or reduced interlacement points are responsible for satin fabric's natural luster owing to more light reflection from fabric surface.

However, the drawback associated with fewer number of interlacements is the poor abrasion of satin woven fabrics especially after repeated wearings. Delicate fashion accessories like stoles, scarves, hair accessories and ornamental clutch, slingbags are designed using satin fabrics.

Pile weave is produced by three sets of yarns, warp, filling, and an additional set that may be either warps or fillings which constitutes the dimensional surface. This woven structure is used to obtain a furlike, plush surface where the raised surface yarns may be left undisturbed like in , terry-cloth, or may be cut as in velvet. Terry cloth has limited applications in accessory industry however velvet is extensively used for luxurious accessories like high-end clutch, potli bags, scarves, hats, hair accessories, etc.

The most sophisticated and complicated weaving technique used for the woven fabric production is jacquard weaving. Conventional methods deployed to produce intricate patterns on jacquard looms used a series of punched cards. However, the advent of technology introduced computer aided designing concept in jacquard weaving which enabled quick and efficient reproduction of patterns unlike the tedious and time consuming setup by the use of punched cards. Jacquard woven fabrics are expensive compared to other woven structures which is attributed to exclusive, intricate jacquard motifs and complicated manufacturing techniques. Accordingly, the jacquard woven fabrics are utilized for designing exorbitant accessories featuring unique motifs and intricate patterns like banarasi scarves, stoles, luxurious handbags, clutches, etc.

Dobby weaving technique is used to obtain simple geometric patterns. Birdseye pique is one variant of dobby fabric which is extensively utilized in females' sportswear accessories, men's ties, etc.

Cotton fabrics are the major raw materials for apparels and accessories. Broadcloth, chambray, corduroy, denim, oxford, plisse, poplin, sateen and seer-sucker are some varieties of cotton fabrics being used for such applications.

Broadcloth is a plain-weave, smooth, flat fabric generally finding application in shirts, blouses and sportswear. Chambray at large is used for nightwear, kidswear, sports apparels and accessories. It is a plain-weave fabric produced using a set of white and colored yarns. Another type of cotton fabric that is extensively utilized for trousers, jackets and other casual wear is Corduroy fabric. Corduroy is a ribbed, pile fabric available in a variety of patterns and weights.

Denim is a twill woven fabric characterized for its durability and rugged appearance and used for designing a range of apparels and accessories like pair of denims, jackets, outerwear, leisurewear, handbags, footwear, hats, etc. Oxford fabrics are plain or basket woven fabrics that are widely used for formal and casual shirts. Plisse is cotton fabric with crinkled appearance due to application of caustic soda and can be used for ladieswear, scarves, stoles, etc. Casual apparels and accessories intended for summer wear generally uses a finely ribbed fabric, poplin. Apparels and accessories that require smooth, lustrous texture uses satin fabric as the base materials. Seer-sucker is another variety of cotton fabric that portrays a crinkled, bumpy texture. The unique texture of the fabric makes it suitable for summer wear accessories, suits and shirts.

Another commodity fibre, flax has been successful in making its mark in apparel and accessory industry owing to its salient, unique features. Flax is obtained from stems of straw like plant and is usually gold colored in its natural state. Linen fabrics, a favourite choice for shirts and other light-weight, summer accessories is produced from flax fibre. The desirable characteristics of flax that makes it suitable for apparels and accessories include its durability, easy wash care properties, moisture absorption and rapid drying, silky, lustrous look and comfort aspects. However, drawback associated with this fibre is its wrinkling tendency and difficulty in pressing.

Art linen, cambric, canvas, crash coarse, damask, handkerchief linen and lawn are different types of linen fabrics used in apparel and accessories.

Art linen is closely woven fabric used for embroidered apparels and accessories. Cambric is a fine linen generally used in handkerchiefs, pocket squares and ties. Canvas is a tough, coarse linen fabrics serving as main constituent for tote handbags, sneakers, baseball caps, etc. Damask is a jacquard woven fabric that is characterized by intricate patterns and three-dimensional appearance with varying shiny and dull surfaces. The fabric finds application in home textiles like table linens, bed linens however, used in apparels and accessory construction to a limited extent. Handkerchief linen also referred to as cambric is used in handkerchiefs and blouses.

While lawn is the sheerest, finest linen fabric generally used for delicate blouses.

Woolens and worsted are the main classifications for wool yarns and fabrics. Woolen fabrics are produced from carded or combed short fibres and exhibit fuzziness, thickness and bulk. Worsted fabrics are distinct from wool fabric characterizing crispness, smoothness, and lighter in weight than woolens.

The innumerous advantages offered by wool fabrics are their durability, snag and tear resistance and anti pill properties. Moreover, wool fibre fabrics are resilient, resists wrinkling, drapes beautifully, and maintains resiliency in wet state. It absorbs moisture, holds air layers next to skin providing warmth in winter and coolness in summer, sheds liquids and breathes, making it comfortable to wear.

Adaptability – It can be used in fabrics that range from the most sheer to the most dense.

Performance – It dry cleans easily, retaining its shape, and can be steamed or pressed with little effort.

Some of the drawbacks to the fibre are its inability to be laundered, except if specially treated, shrinkage is considerable and if not mothproofed, it is an easy victim of moth.

Some of the popular wool fabrics are:

Convert – A closely woven, light weight twill fabric that is used for top-coats, suits and sportswear.

Crepe – A fabric that has a pebbly or crinkled surface that comes in various weights and is used for dresses, blouses, coats and suits.

Donegal – Woolen tweed fabric that is characterized by randomly woven multi-colored yarns.

Felt – A non-woven fabric that is primarily used for hats and trimmings.

Flannel – Lightweight, soft cloth that has a slightly napped finish.

Gabradine – A durable, tightly woven worsted fabric that has a hard finish and is used for men's and women's apparels.

Herringbone – A twill woven fabric that produces a zigzag design.

Loden – A thick, coarse, coating material.

Merino – A fine grade of wool from merino sheep. It is excellent for high quality woven and knitted materials.

Serge – A durable, twill woven fabric that is used for men's and women's clothing

Shetland – A soft, raised fabric made from wool from the sheep in the Shetland Islands in Scotland.

Tweed – A rough surface fabric that uses a variety of colors to achieve various patterns.

Silk – Silk is a filament or continuous fibre a silkworm spins for its cocoon. Unlike the other natural fibres, which come in short or staple lengths and must be attached or spun to form a yarn, a single silk filament ranges from 800 to 1300 yards in length.

Some of silk's advantages are:

Strength – Because it is the strongest of the natural fibres, it resists abrasion and gives the wearer many years of service.

Comfort – It is extremely smooth to the touch and is soft on the skin. It is elastic and retains its original shape. As it readily absorbs moisture, it is comfortable in warm atmosphere.

Care – It may be dry cleaned or washed. Washing however, should be done by hand, unless the garment has been processed for machine laundered.

Shrinkage – The amount of shrinkage is minimal.

Appearance – It is the most luxurious of all fibres, with a natural luster unknown to other fibres.

Silk tends to water spot, will damage when exposed to perspiration, and will weaken if continuously exposed to sunlight. Silk comes in a variety of weights and textures. Some of those used in apparels and accessories include:

Alpaca – A fine-textured, slightly lustrous fabric that is woven to resemble cloth made from South Americans alpacas.

Broadcloth – A spun silk, woven with a firm yet soft feel.

Boucle – A nubby textured fabric, woven of yarn that has been twisted to produce a loopy thread.

Chiffon – An airy, gossamer sheer that has a floating quality in motion.

Crepe de China - A lustrous, light crepe with a soft rippled effect.

Geogrette – A soft, sheer fabric with a crepelike texture.

Satin – A smooth, shining fabric.

Shantung – A soft, lustrous fabric with sparse, irregular ridges in the texture.

Taffeta – A fine, lustrous fabric that is characterized by both its crispness and sound made through movement.

Figure 3.9 Woven fabrics

Pongee – A light or medium weight Chinese silk fabric made from Tussah silk.

Tussah – A springy silk with sparse, irregular ridges in the texture that has been woven from silk of uncultivated silkworms in Asia.

3.1.3.2 Knitted fabrics

Knitted fabrics are produced by interlocking loop formed by continuous yarn strands that are pulled through one another in loop configuration by hooked needles. The process basically involves pulling a newly formed loop through the previously formed one thus producing the desired knit structure.

Hand knitted as well as machine knitted fabrics both find applications in apparel and accessory industry. The former, however, is costly and has a limited market while the latter is extensively utilized to meet the ever-increasing demands of knitted goods in domestic as well as international market.

Knitted fabrics are classified into weft and warp knitted fabrics. Weft knit structures namely single jersey, rib and interlock are produced horizontally with the stitches running from side to side along the fabric width. Single jersey with lycra, single lacoste, double lacoste, single pique, double pique, polo pique, French pique, terry with lycra,1*1 Rib, 2*1 rib, 2*2 rib, lycra rib, flat back rib, drop needle interlock and interlock with lycra are some variants of weft knitted fabrics that vary considerably in their structures and properties and widely used for a range of apparels and accessories. Warp knitted fabric such as tricot and raschel are produced utilizing a large number of yarns with each yarn looped around a single needle and attaching vertically to other loops.

Knitted fabrics are quite distinct from woven fabric as far as dimensional stability, flexibility and stretchability are concerned. The knits are dimensionally less stable compared to wovens which may be attributed to interlooping of yarn whereby each newly formed loop is dependent on the previous one. Consequently, breakage or damage to one loop in the structure leads to complete structural breakdown referred to as laddering. The low dimensional stability and easy unravelling of knits restrict their usage for apparels and accessories that demand dimensional stability and strength. However, knitted fabrics outshine woven fabrics in terms of stretchability, porosity, better air entrapment and improved thermal insulation combined with good moisture absorption and comfort aspects ensuring mobility and conforming fit to the wearer. The icing on the cake is less production costs associated with knitted fabrics compared to their woven counterparts. Accordingly, knitted fabrics are considered to be ideal choice for winter wear accessories like socks, stockings, gloves, hosiery requiring a close fit with no bagging, stoles, mittens, veils, mufflers and beanies (a famous winter hat style) owing to inherent ability of knits to entrap large volume of still air, retaining the body heat and thus providing effective thermal insulation.

Furthermore, knitted fabrics using fibres like polyester, nylon and spandex and mesh structures are also being used for footwear designing as the structures show the capability of providing an additional degree of comfort stretch and are available in lower price range unlike the leather footwear.

Figure 3.10 Knitted fabrics

3.1.3.3 Non-woven fabric

Non-wovens are distinct from woven and knitted fabrics in terms of their structure and production methods. Non-woven fabrics are constituted of a fibrous web of randomly arranged fibres, bonded together by chemical, mechanical, heat or solvent bonding techniques. Film coated fabric closely resembling leather is also a non-woven material. Non-wovens are versatile fabrics offering high standards of quality, softness and resilience, exhibiting a range of properties like absorbency, liquid repellence, resilience, stretch, softness, strength, flame retardancy, washability, cushioning, thermal insulation, acoustic insulation and filtration. Accordingly, the application areas of non-woven fabrics are vast spanning from civil engineering, road construction to different textile segments. Non-woven fabrics are used in a gamut of commodity and functional apparels and accessories as well as medical textiles, home textiles, technical textiles, etc. Gloves & glove liners, medical & surgical apparels and accessories like surgeons' gowns, coveralls, gloves, head and foot covers, bandages, sanitary or anti-bacterial wipes, diapers, adult incontinence prevention and feminine hygiene products, military apparels, outerwear, sportswear & swimwear, protective clothing & accessories like chemical protective gloves and face masks, caps, fire-fighting gloves, shoe linings & insoles, sleepwear, carpeting, upholstery, shoulder pads interlinings and interfacings are some of the application areas of non-wovens.

Interlinings, felt, vinyl and suede cloth belong to non-woven category and are vital raw materials in fashion industry. Non-woven interlinings are widely used as a reinforcement and stabilizing medium in garments constructed from sheer fabrics or those garment portions which are prone to abrasion like collars, cuffs and plackets. Furthermore, non-woven interlinings are also used as backing materials in accessories like handbags, hats and footwear, etc.

Felted fabric that belongs to category of non-woven is produced by felting process whereby wool fibres are directly transformed into fabrics omitting the intermediate yarn formation process. Felted fabric production requires a series of mechanical operations like agitation, friction, and pressure application in the presence of heat and moisture. The fibres are prevented from recovering to their original position due to interlocking of scales on the fibre edges during mechanical action.

Felt fabric can be easily cut without any fraying and molded into desired shape with an additional advantage of superior insulation properties which can be accounted to densely packed fibres in the structure. Felt fabric owing to its low dimensional stability is pertinent choice for fashion accessories that demand minimal laundering, pulling, or stretching. Henceforth, felt fabric is extensively used for making accessories like structured hats like fedora, cowboy, derby and other winter wear accessories which benefit from dense structure and warmth properties of felt. Likewise, modacrylic fibre felts are being widely used in designing footwear and handbags for fall/winter seasons. However, one disadvantage associated with felt fabric is its stiffness making it a difficult fabric to drape into graceful folds.

Vinyl is made by coating base fabric with a polyvinyl chloride (PVC) or polyurethane film resulting in a non-woven structure resembling patent leather. Handbag, footwear and hat designers attempt to create ornate accessories from embossed vinyl that has been surface treated to imitate skin of rare, expensive animal species.

Suede cloth used for designing exclusive fall and winter accessories like espadrilles, loafers, sneakers (footwear styles), handbags and hair accessories, is non-wovens web constituted of randomly arranged microfibres.

3.1.3.4 Miscellaneous fabrics for accessory construction

(a) Macrame

Macrame meaning napkin or towel in Turkish, is a technique that resembles geometric weaving process used for net making. The technique involves looping and knotting of the cords with beads inserted strategically so as to offer patterned designs combined with ornamental knots. Fashion accessories such as hair bands, tiara, necklaces, earrings, bracelets, handbags, belts and fashion footwear like espadrille, thongs, etc. made by macrame technique can be aesthetically ornamented by beads and cords of varying colors, sizes and textures.

(b) Crocheted fabrics

Crocheted fabrics are formed by looping and interlocking of the yarns employing a single hooked needle. Crocheting produces the most exclusive and intricate lace patterns accomplished either by hand or machine. The technique renders scalloped edging on innumerous accessories like handkerchiefs, hats, beanies, hair accessories, neckpieces and earrings, handbags, shawls and stoles for edge neatening and ornamentation. Moreover, apart from edge neatening, crocheted fabric can serve as base fabric for construction of delicate handbags like tote, satchel, slingbag, hats and gloves.

Figure 3.11 Crocheted fashion accessories

(c) Braided fabric

Braided fabric is a narrow woven structure formed by intertwining and twisting of three or more strands of fibres, yarns or threads with each other in some angular configuration. Braiding is classified as two and three-dimensional braiding. The former can be circular or flat braid while the latter is a relatively new technique and utilized for composite structures.

Although braided structures are more significant for industrial fabrics, harnesses, hoses, industrial belts and surgical sutures than consumer textiles however, accessory industry utilizes braided fabrics, laces and trimmings in multitude applications like wristbands composed of several colored yarns, threads or twines in braided arrangement, braided watch bands, head bands and hair accessories, belts, footwear, golf bags and even as decorative edging and neatening on scarves, stoles, gloves, mittens, etc. As far as apparel industry is concerned, braided fabrics are widely used as ribbon tapes, plain tapes, jacquard Tapes in men's trunk tops, ladies lingerie, sports and active wears like tennis shorts, jogging suits, swimwear, etc.

Figure 3.12 Macrame & braiding techniques for designing fashion accessories

(d) Quilted fabrics

Apart form woven, knitted and non-woven fabrics, quilted, coated and bonded fabrics are also used for apparel and accessory designing.

Quilted fabrics are composed of three components namely a cotton batting, wool or down filling material and two layers of decorative outer fabrics. The batting is sandwiched between the two decorative layers with top layer constructed of small fabric patches pieced together . The filling material is kept in place between two outer layers by stitching the layers of the quilt together. Stitches are located in such a way so as to create a geometric, floral padded design. Quilted fabrics are preferred for fleece winter jackets and accessories like gloves, handbags, footwear, etc.

(e) Bonded fabrics

Bonded fabrics are produced as a result of adherence of two fabrics having similar or variable composition and construction. The front or face side of bonded fabric is generally woven while knitted fabrics are used as the backing material. Adherence of webs to two layers result in the production of bonded fabrics. The fabrics find application in footwear and handbags, etc.

3.2 Non-textile materials

The previous section of chapter discussed the various textile raw materials used in accessory designing. However, the contribution of non-textile raw materials in apparel and accessory industry can not be undermined. This section will elaborately cover the various non-textile materials being used in apparel and accessory industry. The non-textile raw materials generally finding usage in accessory industry can be broadly named as leather, fur, metals and stones.

3.2.1 Leather

Leather is the generic term for all hides and skins obtained from land animals like cattle, sheep and goats. Leather encompasses all of the processed skins and hides (fur removed) from land animals, reptiles, fish and birds (feathers removed). The hair present on animal skin is generally removed to be used independently as fur or down feather.

Putrefaction is the decay of animal hide or skin caused by microbial growth on the skin which causes discoloration or yellowing of the skin, generating bad odour and making it unsuitable for textile applications. So, the animal skin needs to be preserved by processes like curing and tanning so that putrification can be prevented.

Hides are pelts obtained from larger animals such as cattle, horses, alligators, etc. while that obtained from skins of smaller animals like calves, sheep and goats is termed as skin.

These animals are usually slaughtered specifically for their skins. The animal groups that are mainly reared for their hides and skin to be ultimately processed into leather are cattle, cow, bulls, buffalo, sheep, lambs, goat, equine horse, zebra, pig, reindeer, kangaroo and ostrich, etc. Exotic leather is obtained from a range of amphibians, aquatic as well as land animals. Frog, seal, shark, walrus, and turtles are the aquatic of the group, camel, ostrich, elephant and pangolin are the land classification, and alligator, crocodile, lizard, and snake represent the reptiles which are used for leather manufacturing. Each has distinctive characteristics and markings that make them appropriate for unusual apparel and accessories. Thickness of the skin, durability, leather processing and rarity are some of the parameters that gauge the suitability of a particular leather variety for a specific end-use.

Although, a variety of land and aquatic animals serve as leather source however, the most common source of leather – bovine leather, by far is cattle industry. Apart from bovine leather which is undoubtedly a preferred choice for apparel and accessory designers, imitation or simulated leather has been doing rounds and gaining popularity among fashion buffs. Leatherette , naugahyde, vinyl, pleather, liquid leather, super-suede, facile and ultrasuede are the varieties of simulated leather that are specially treated to acquire washability and durability matching that of real leather. Woven or knitted base fabric coated with polyvinyl chloride serves as the main constituents for water proof simulated leather. The base fabric provides stabilization, while the coating or film makes the leather resistant to dirt, chemicals and abrasion. Needle punched fabrics composed of polyester and polyurethane foam are being used to create suede-like material. The detailed description of leather types will follow in the later part of chapter.

Cheaper substitutes of leather aimed at cost cutting prompts the creation of simulated, imitated leather by shredding and reconstitution of leather scraps, however it can not match the durability associated with patent leather. Shredded and reconstituted version of leather is sold as labeled ground leather, pulverized leather, shredded leather, reconstituted leather or bonded leather.

Leather both patent, virgin as well as imitated version like suede, bonded leather holds lion's share of market as far as footwear, belts, wristbands, watch straps, gloves, luggage, handbags, wallets, etc. are concerned.

Outsoles and heels of footwear like thongs and slipper, briefcases, gloves, handbags, and wallets utilize cattle steer, cow and bull hides in their construction. Likewise, shoe uppers and linings, gloves and hats generally use kipskins or kips obtained from large calves or undersized cattle. The woolen skins, hair skins, and shearlings obtained from sheep and lamb are used for designing accessories like moccasin, oxford (footwear styles), biking gloves, slingbags, clutch, tote bags, derby and cowboy hats.

Figure shows a range of leather accessories like handbags, laptop and sling bags, wallets, mobile covers, clutch, belts, gloves, footwear, hats, baseball caps, jewelry and hair accessories.

Different Parts of Leather Hide

Cowhide, one of the most common source of leather, is quite thick, and is usually cut down to be thinner as per the end-use requirement. Cowhide is composed of two integrated layers – the corium and the grain (Figure 3.13). Corium layer consists of thinner and more flexible collagen fibres which become tighter and thicker on upper side toward the grain. The fibres in upper layer are tightly packed and sturdy. The corium gets thicker with animal's age and hence hides of older animals are thicker compared to calfskins which are characterized by their fineness, smoothness and softer feel.

The grain of the skin or hide is the uppermost skin surface from which the hair grows. Split is the layer below grain. The full or top grain is the layer of the skin with inherent, characteristic markings prominent as the hair or fur fibres have been removed.

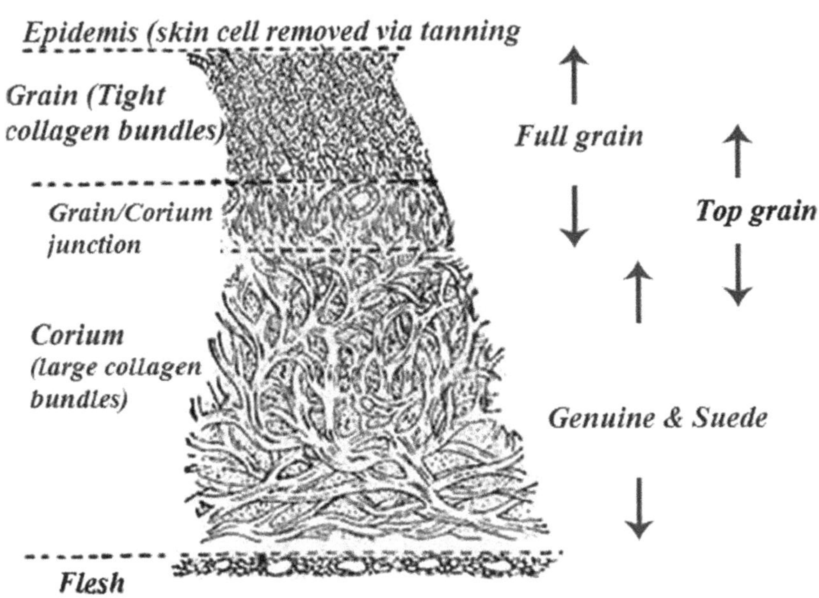

Figure 3.13 Different parts of leather hide

The top part of the grain which faces outward toward the hair may include blemishes like insect bites, stretch marks, scars, etc. which can impair the aesthetic appeal and texture of leather. Hence, the top part of grain requires buffing to render it a more uniform look.

Top grain requires some correction to remove blemishes however, full grain leather has entire grain intact, with some visible blemishes as it does not receive any sanding or buffing treatment. Full grain and top grain leathers together are referred to as grain leather.

Full Grain or Top Grain Leather is obtained from the upper layer of a hide which undergoes splitting into layers of different thickness with the aid of splitting machine.

The split leather is obtained by splitting off the bottom layer from the grain at the grain/corium junction.

Figure 3.14 Leather accessories

Properties of leather suitable for apparels & accessories

Leather has several unique and salient properties that make it acceptable as raw material for a range of apparels and accessories.

(a) Tensile & tear strength leather possesses extremely high tensile strength, a much sought after property for designing accessories that demand rough and tough usage, durability and substantial load carrying capacity. It can withstand significant amount of stress without tearing apart. Leather further outshines other textile materials in terms of tearing strength and can effectively resist tearing.

(b) Elongation & flexibility leather exhibits elongation and can be stretched to varying lengths without any structural breakdown. Moreover, it shows flexibility over a wide range of temperature. Brittness and cracking of leather in extremes of climates can be avoided by tanning or preservation.

(c) Absorption leather exhibits moisture and liquid absorption property which is a crucial property for footwear for absorption of perspiration from the lining materials and rapid spreading throughout the leather uppers to ensure foot comfort.

(d) Breathability and insulation leather outerwear and accessories can be worn in both cool or warm climates as it can adjust to the outside temperatures due to its breathability and insulation properties imparted during processing.

(e) Lasting moldability leather has the ability to conform and retain its shape during the service life of accessories like footwear, hats which undergo lasting process whereby leather shoe upper or crown of hats are pulled over or forced onto last (wooden replica of footwear, hats).

Leather processing

Leather is a major raw material for a range of apparels and accessories. Leather should be imparted textile like properties like moisture absorption, softness and comfort next to skin to be used as a wearable end product. The hide obtained in raw form from animals is not suitable for next to skin and other textile applications. Consequently, the hides or skins are subjected to a series of operations to convert it into usable and wearable leather. The leather manufacturing process can be subdivided into four main processing stages – beamhouse operations, tanning, post-tanning and finishing.

Beamhouse operations

Beamhouse operations encompass the preparation of the hides for tanning. A number of undesirable components like any foreign dust, dirt particles, animal blood present on the hide should be removed in preliminary steps prior to tanning. The preparation of pelt (hide devoid of hair fibres) prior to tanning and involving the processes from soaking to pickling are collectively referred to as beamhouse operations.

Curing The skins in raw form obtained from animals is prone to rapid putrefaction and requires immediate attention for its preservation after the animals are slaughtering off and before being sent to the tanneries for further processing. Curing is the initial preservation process to preserve raw hides and skins and prevent them from putrefaction, yellowing, rottening, bad odour generation before the beginning of leather-making process. Hides and skins are preserved using salt to obtain a dehydrated wet-salted or dry-salted material. Apart from salting, the hides can be preserved by chilling, freezing and the use of biocides.

Thereafter, the raw hides are checked for quality, any irregularity in color, thickness and sorted by weight **before** sending them to tanneries.

Soaking

Curing is followed by soaking when the cured hides are soaked in water for duration ranging from several hours to several days. The hides dehydrated during curing must be rehydrated before it it is subjected to any further mechanical action. Water is not only essential for chemical penetration and removal of salt, dirt, etc. but is also required for all the subsequent processes. Furthermore, soaking assists in the removal of curing agents, non-structural proteins and fats. The hide is prepared for further processes and enables it to reabsorb any water lost in the curing process or during transportation and free it of any salt, dirt, etc. Soaking makes the hides swell and provide them softer feel. The purpose of soaking can be broadly stated as:

- Restoration of the original water content in the hides after curing.
- Swelling of the hides
- Removal of any foreign particles, dirt, manure and blood.
- Removal of preservatives and disinfectants.

Unhairing and Liming

The clean hides obtained after soaking are further subjected to liming process which is one of the crucial steps in beamhouse operations. The primary function of liming is the removal of any unwanted hair and epidermis from the hide by using alkali. Raw animal skin obtained after the hair removal is referred to as pelt. Pelt, the hairless hide is very greasy owing to its high fat content. The alkali used during liming causes swelling of the pelt and controlled breakdown of some of the chemical crosslinks of the collagen. The removal of hair or wool fibres from the hide is accomplished by soaking in liming drums containing sodium sulphide and hydrated lime. The keratinous material of the epidermis and hair or wool are destroyed on exposure to the chemicals thereby revealing the distinctive pattern of the grain surface. A turgid, open-structured, translucent, jelly-like material is formed due to hydrolysis of fats and swelling of the skin structure.

The objectives of liming can thus be summarized as:

- Easy removal of hair by loosening the top skin and hair
- Removal of meat and fat residues by loosening
- Removal of fat in the hides.
- Swelling of the leather fibres.
- Opening the skin structure, making it receptive to chemicals in further processes.

Fleshing

Liming is followed by fleshing where a special machine removes fleshy tissue from the flesh side and splits the pelt into layers. Fleshing removes the connective tissue and meat residue from the flesh side of the skin. The hides are in a soaked yet softened condition till this stage. The older methods of flesh removal used sharp, curved blade to remove excess flesh from hides which were placed on a beam. More recently, rotating scraping rolls are being used to remove flesh from the hides. The debris remaining after flesh removal is referred to as glue stock.

Splitting

Spilitting involves cutting the hide/skin into two or more horizontal layers to obtain uniform thickness. The top layer, or grain side produces a fine, smooth grain leather while the bottom layer is transformed into suede or split leather.

Deliming & Bating

Deliming is carried out on split hides to prevent the distortion or disruption of the tissues in the pelt. The process is accomplished by neutralising the alkali using weak acids thereby avoiding any rapid changes in pH. Deliming is followed by bating in which hides are treated with proteolytic bating enzymes, thus cleaning the grain and providing a smooth and silky appearance to the pelt. The enzymes used during bating are similar to those found in the digestive system that aid in breaking down proteins and imparting desired softness to the leather The duration of the bating process depends largely on the softness required in the leather. Longer baiting process will result in a leather with enhanced softness.

Pickling

Pickling process is carried out on bated hides to make them receptive to the tanning agents used during tanning. The optimal penetration of the tanning agent require control over pH of the hide and accordingly, the hides are treated with acid (sulphuric and/or formic acid). The acid application maintains the hide at pH of 3 enabling chromium tannins to enter the hide. The fibre swelling is avoided by addition of salts. Further, fungicides and bactericides are also used. For preservation of pelts (hide with hair removed). The so treated pelt (pickled pelt) after this stage, is ready for a pretannage operations.

Pretanning, tanning & retanning

Pretanning

The penetration and distribution of tanning agents can further be improved by preliminary light tanning prior to the main tannage. Moreover, pretanning assists in stabilising the leather for any further mechanical operations like shaving.

Tanning

Tanning is the process of preserving animal skin making it resistant to microbial attack and prevention from decaying, decomposing and oxidising. The process is carried out in tannery where the hides and skins of slaughtered animals are treated with tanning agents for irreversible chemical preservation and conversion into usable leather. The animal skin is composed of one third collagen protein. The purpose of adding tannins to hide/skin is to retain the flexibility of protein strands and prevention of hardening and sticking of the protein filaments on drying. The process converts the protein content of the raw hide into a stable material preventing it from putrefaction and making it suitable for a range of textile applications. Tanning agents form crosslinks in the collagen structure and stabilise it against the effects of acids, alkalis, heat, water and the action of micro-organisms.

The operating steps in tanneries subject the leather to rotation in barrels irrespective of the tanning agents employed for tanning. Drums, paddles, pits and mixers are some process vessels which are used for tanning. Wooden, stainless steel and plastic drums are generally used for tanning. However, wooden drums owing to their ability to withstand strong alkaline solutions and chemicals used in the initial stages of the tanning processes are preferred over the other two drum types. Moreover, wooden drums used in tanneries are sturdy and possess good insulation properties, unlike metal. The lower costs, easy maintenance and repairs as compared to stainless steel drums also makes wooden drums the first choice among tanners.

Purpose of tanning

Tanning in leather imparts several desirable properties to hide and thus converting it into usable, wearable leather. Tanning is performed to incorporate following properties in leather:

- Imparting resistance to leather against decaying, tearing, abrasion, light
- Prevention of fibre swelling in wet leather
- Increased temperature resistance of leather. Untanned skin shows shrinkage in water at 62°C while chrome-tanned leather is resistant up to 100°C.
- Avoiding fibre sticking and shrinkage of leather
- Imparting softness or firmness to leather
- Imparting durability and breathability to leather
- Obtaining the desirable colour and gloss level
- Improving after use care and handling properties to leather

Different methods employed for tanning

Tanning can be categorized into real and false tanning based on the type of bonding achieved between tanning agents and collagen structure of the hide and by the colour imparted to leather by different tannins.

Chromium tanning salts, vegetable tannins and glutaraldehyde are some of the tanning agents used for tanning. Real tanning is referred to as a tanning process in which tannins form an irreversible bond with collagen fibres of the hide as in case of chrome tanning and vegetable tanning processes.

The penetration and distribution of tanning agents into the collagenous structure of hides causes irreversible binding of tanning agents to the collagen via increase in pH. The process of permanent, irreversible binding of chrome, vegetable or any other tanning agents to collagenous hides is referred to as basification.

Chrome tanned leathers tanned with mineral based tanning agents like salts of chromium are comparatively less desirable than leather tanned with other agents because of their mineral content of the former. More recently, leather is tanned with inert chrome III which is an essential dietary element that causes metabolism of normal glucose, protein and fat.

Tanning leather by use of vegetable tanning agents obtained from plant extracts result in thick , firm, brown coloured leathers. Vegetable tanned leathers are suitable for making belts, shoe linings, bags and wallets, mobile and spectacle cases.

A specific desirable property in leather can be achieved by using various tanning methods in combination with one another, referred to as combination tanning, e.g. combination of synthetic tanning with chrome or vegetable tanning. Combination of vegetable and subsequent chrome tanning results in semi-chrome leather while chrome tanning followed by vegetable tanning is accomplished during chrome retanning.

False tanning is the term used for tanning process where binding of tannins to the leather fibre is of temporary nature, washable and unstable. Tawing, tanning with fats and oils are false tanning methods with reversible binding between collagenous structure and tanning agents.

The most conventional method of tanning that has been used for ages in tanneries is Tawing. Tawing produces white leather by subjecting animal hide to a mixture of alum (aluminium sulphate) and saline. However, the white leather obtained by this method is not washable as the process is not permanent and the tannins can be easily flushed out by water.

Another tanning method employed in tanneries is by application of fats and oils. Hence, fatty animal substances like brain, fish oil or tallow are used as tanning agents in this method and the leather so produced is referred to as chamois leather.

Aldehyde and oil based tanning agents produce soft leathers that are dry cleanable or easily washable. Treatment of dried raw hide with oils result in formation of parchment which cannot be regarded as leather. The aldehyde tanned leathers fulfill the requirement of automotive sector however, handling, effluent treatment and higher energy consumption can be issues related to aldehyde tanners.

Different tanning methods are also distinguishable based on the color imparted to leather before colouring and finishing. Vegetable tanning agents impart brown color to leather while tawing produces a white colored leather. Chrome tanned leather displays bluish grey (wet blue) color tone.

Retanning

Retanning is a mandatory step performed on tanned leather to achieve the desired characteristics like softer, suppler feel, dyeability, fullness, fineness and stability of the grain, light fastness and other properties dictated by leather apparel and accessory designers and manufacturers. The tannins used for retanning are different from those utilized in the initial tanning stage.

Re-tanning imparts faded effect to leather which can further be enhanced by application of lightening tannins. Re-tanning ensures that leather can be bleached to be dyed indifferent shades subsequently. The process is accomplished by addition of pigments to drum that results in surface colouration of the retanned leather.

Neutralising

Neutralising is essential post tanning to remove any leftover acids from tanning. The process uses alkaline sodium carbonate with concentration of 1.2% to maintain a pH value of 4-6 which allow subsequent process of fatliquoring to occur smoothly. Higher pH value of leather in the range of 7 at this stage may dissolve binding of the tannins to the collagen fibres of hides. Consequently a drastic reduction in durability of leather will be encountered.

Neutralisation

The mechanical operations performed on leather usually squeezes out water from leather, thus a wetting back and washing process is required prior to further treatments to rehydrate the leather and removal of dirt, shavings or grease. Most of the chemicals to be added in further processes are anionic in nature. However, the tanned collagen at low pH is cationic in nature. It thus becomes essential to neutralize leather by raising pH and addition of chemicals that reduce the harshness of anionic chemicals such as retanning agents, dyes, etc. on leather.

Withering

Withering is the next operation performed on neutralized skins to squeeze out residual water from wet skins by forcing them through metal rolls.

Several methods are employed for drying the leather in the intermediate steps of the leather production. Leather generally requires drying twice between the various process stages. The drying methods commonly employed in tanneries include hanging, drying tunnels, drying machines, high-frequency drying, drying in the sun, etc.

The drying process selected depends on time allowed for drying. Shorter drying times result in higher energy costs. A consideration of impact of drying method on the size of the surface and smoothness of the leather is thus crucial.

Mechanical dewatering

Mechanical dewatering makes use of special roll-presses for pressing damp, freshly tanned hides between two metal cylinders to flush out the excess water. The felt covered rollers are used to avoid folding of hides as they are passed through the rollers. The process is time and energy efficient and comparable to wringing damp garments before hanging for drying in sunlight. Water content in the hides is reduced by about 35 to 50% by mechanical dewatering.

Air –dried hides/ ground drying/frame drying

Hides can also be dried by exposing the flesh side to mild sun rays. The process is undoubtedly cheaper but also a time consuming one. The hides are flat spread on the floor or can be stretched in tensed state onto a square wooden frame. The stretched hides placed in drying frames occupy more surface and show less shrinkage during the drying process.

Drying by hanging

Leather can also be dried by hanging on hooks or over bars at normal room temperature by increasing air movement. The process may take several weeks to dry leather depending on the type of leather. Drying by hanging is suitable for thin and soft leathers. However, leather undergoes deformation due to loose hanging without any supporting frame.

Hot air drying in drying tunnels

The hides are dried by placing them on a bed composed of perforated metal plates and keeping in stretched condition by toggle clips. The metal plates are exposed to circulating hot air as they are move through drying tunnels. The method ensures drying hides in less time as compared to drying by hanging.

Vacuum drying

Vacuum drying is one of the preferred methods for drying leather where leather is positioned with the grain side facing the heated polished metal surfaces. The drying equipment consists of vacuum chamber through which water vapour is sucked out as leather is dried. This drying process is based on the physical principle of water evaporation at low pressure and low temperature and is advantageous in causing liquid to flow through the flesh side and drawing greases from the grain side into the inner zone of the hide. The process is fast as drying time is short and eliminates most of the manual work as the hides are positioned onto metal plate without clamps or adhesives.

Vacuum drying assists drying of leather in smooth and flat condition unlike the drying of leather in the drying tunnel. The grain side of leather can further be smoothed by ironing. The grain gets stretched and the structure becomes more open on drying leather by stretching onto frames. The frame-dried leather will thus lack the smoothness that can be achieved with vacuum-dried leather.

Glass drying

Glass drying involves drying leather by glueing grain side of moist leather onto glass plates ensuring that leather does not distort or shrink during drying. Wet leathers glued on glass panels, are passed through drying tunnel which has air circulation facility and is maintained at under controlled temperature. The process, thus results in complete drying of leather in few hours providing a smooth leather surface. The advantage of this drying methos is the short drying time and the smooth surface imparted to leather after drying.

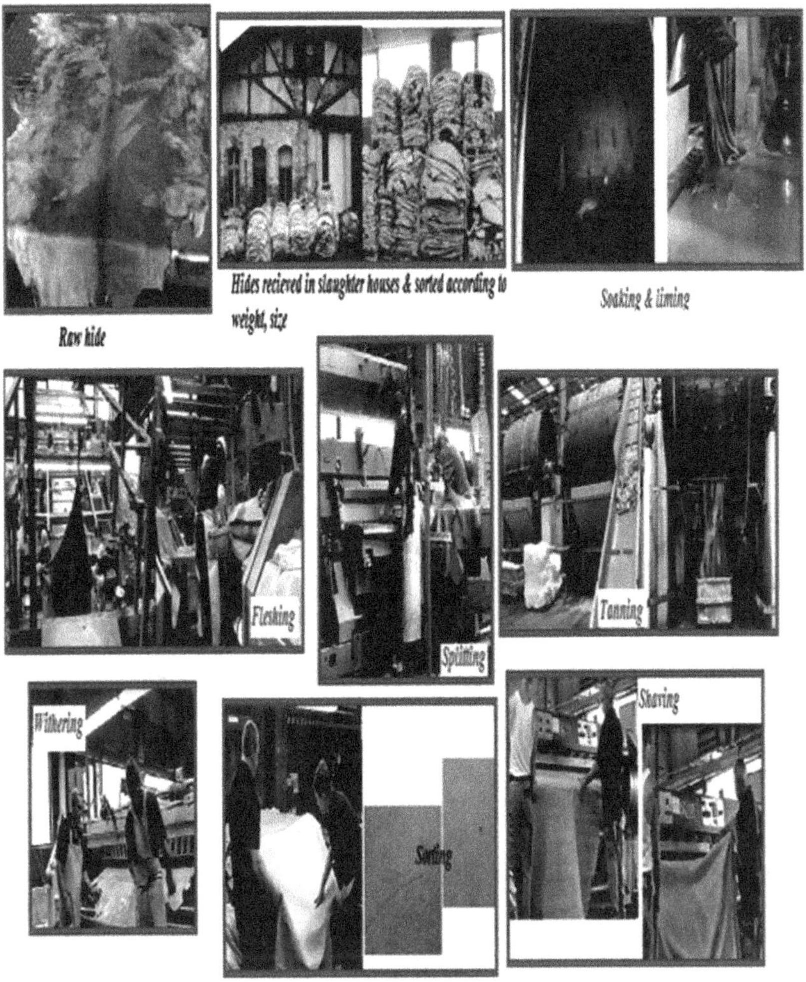

Figure 3.15 Leather Processing

High-frequency drying

High-frequency drying is yet another drying method currently used to limited extent in the leather industry. At initial stage, leather is dewatered by pressing and subsequently humid leather is exposed to an electric field of high-frequencies by a belt conveyor system. Internal heating and water evaporation occurs due to the energy density in the humid, conductive areas.

Drying is followed by tumbling, stacking of leather to achieve the required level of softness in the processed leather.

Tumbling

Tumbling process is carried out to soften leather by subjecting it to rotating movement in a narrow, high, rotating barrel. The barrel contains short rods and steps that reinforce the leather's movement. This process is essential as leather fibres have tendency to stick together and becoming firmer during drying. Moreover, the application of finishes on leather can make it stiffer and thus softening becomes prerequisite at this stage.

Leather can further be softened by another process known as stacking. During stacking, a stacker or a stacking machine consisting of two knobbed plates and opposite recesses is used to soften leather. Leather is pressed into recesses by the knobs by constant vibrations and thus stretching the leather fibres. The total surface of the hide increases considerably and grain patterns are reinforced as a result of stretching and softening imparted to leather during stacking.

Sorting

The leather then undergoes a quality check and is sorted according to the quality. Hides devoid of any blemishes or marks are regarded as the exotic, best quality and are processed into aniline leather as they require minimal surface treatment. However, hides with prominent markings, blemishes will distract the visual appeal of leather and require processing into corrected grains.

Finish- Surface coloration & processing

Surface of leather can further be beautified for value addition by various surface coloration and processing techniques. Colouring, pigmentation, waterproofing, wax dressings and mechanical processes like ironing, embossing are some of the methods employed for surface modification and colouration of leather. The appearance of surface and surface properties of leather largely depend on the finishing method used. The finishes applied on pigmented leather consists of a primer, a colour coat, top coat and a crosslinker used for stabilizing the layers.

The durability, stain resistance and water repellency can be imparted to aniline pre-colored leather by pigmentation process. The process is accomplished by treating leather's surface with a layer or coating of a binder-pigment-mixture. Pigmented or finished leather finished with this protective colour layer is utilized for designing biker and bomber jackets, footwear and handbags.

Aniline leather is smooth layer without pigmentation i.e. coating of protective colour layer. Refined or semi-aniline leather has a very this layer of finishing on its surface. However, leather with a finish layer of more than 0.15 millimetres in thickness is referred to as coated leather. Aniline leather provides warm feel while pigmented leather is colder and firmer and is easier to clean and care as compared to aniline leather. However, the latter type exhibits low breathability due to reduced porosity caused by coating on surface. Pull –up leathers have an oil or wax coating on their surface.

Final check

Final check is essential to review the produced leather's quality and size measurement of the skins.

Mechanical operations

Upper grain split and lower flesh split of required thickness can be obtained as the leather is split longitudinally. Chromium tanned leathers production required this process to be accomplished post basification and samming. The leather is then shaved to obtain a more accurate and even thickness. The final achievable thickness after shaving depends on the application areas for which shaved leather will be used like shoe upper require 1.8 mm thick leather while for garment a thickness of 0.9 mm is just appropriate.

Retanning

The tanned leather is subjected to additional tannages with similar or new tanning materials. These agents may be used to lighten the colour of the leather, to produce a feeling of fullness and to aid in the penetration of dyes. The choice of pretanning, tanning and retanning chemicals is dependent on the properties desired in the final leather, and therefore, on the properties required in the final leather product.

Dyeing

The desired colour in leather can be obtained by dyeing. The dye serves as a base colour for finishing, and the depth of dye penetration and leather colour are of interest to both leather apparel and accessory manufacturers alike.

Dyeing

The dyeing of leather into a wide variety of colours plays an important part in meeting fashion requirements. Some leathers are only surface dyed, while others need completely penetrated dyeings, as is the case with suede leathers.

Fatliquoring

The chrome tanned leather on drying becomes hard and crusty and may not be suitable for textile applications. The fullness, softness and flexibility can be incorporated into leather by application of oil emulsions known as fatliquors.

Drying

Leather batches are generally toggle dried on frames in heated tunnels for four to six hours while vacuum drying is a comparatively faster drying process taking just two to ten minutes for leather drying. Dried leather is further subjected to buffing, conditioning, staking and milling.

Finishing

The surface of leather can be modified and beautified imparting functional or aesthetic appeal by application of resins, pigments, dyes, handle modifiers, fillers and dullers, etc. Spraying, roller-coating, curtain-coating or manual methods are employed for finishing leather surface. smooth or patterned leathers are obtained by heated hydraulic or roller presses. Finishing is thus the last stage in leather manufacturing process following which area of leather is measured and finally sent for dispatch to apparel and accessory manufacturers.

Shaving

A uniform thickness is achieved by shaving the leather on the non-grain side using a machine with a helical blades mounted on a rotating cylinder.

Samming

Samming on leather is accomplished by a large mangle equipped with felt covered rollers. The process brings down the water content in leather to about 55%.

Setting out

Setting out is performed to smoothen the grain side and stretch the leather. Moreover, the process is helpful in reducing the water content to about 40%.

Final drying

Leather is dried retaining 10-20% water content which is essential to prevent leather from becoming brittle and reducing its cracking tendency. The method employed for drying leather has a striking impact on final effects achieved in the finished leather.

Staking and dry drumming

Leather is made softer and flexible by the use of staking machine which is equipped with a rotating drum that tumble the leather rigorously and separates the fibres. Thus, the desired softness can be achieved in leather.

Buffing and Brushing

Mechanical abrasion causes the removal of flesh surface reducing the thickness and thereby resulting in suede effect to the buffed leather. The grain surface can also be buffed in some instances producing a very fine napped surface like nubuck leathers. Buffing is followed by brushing to remove excess dust from leather surface.

Finishing

Finishing renders several desirable properties to finished leather. Finishing not only assists in color uniformity but also ensures that any visible grain defects can be concealed, highlighting the glossy surface. Furthermore, finishing offers protective surface to leather imparting resistance against water, chemicals and abrasion.

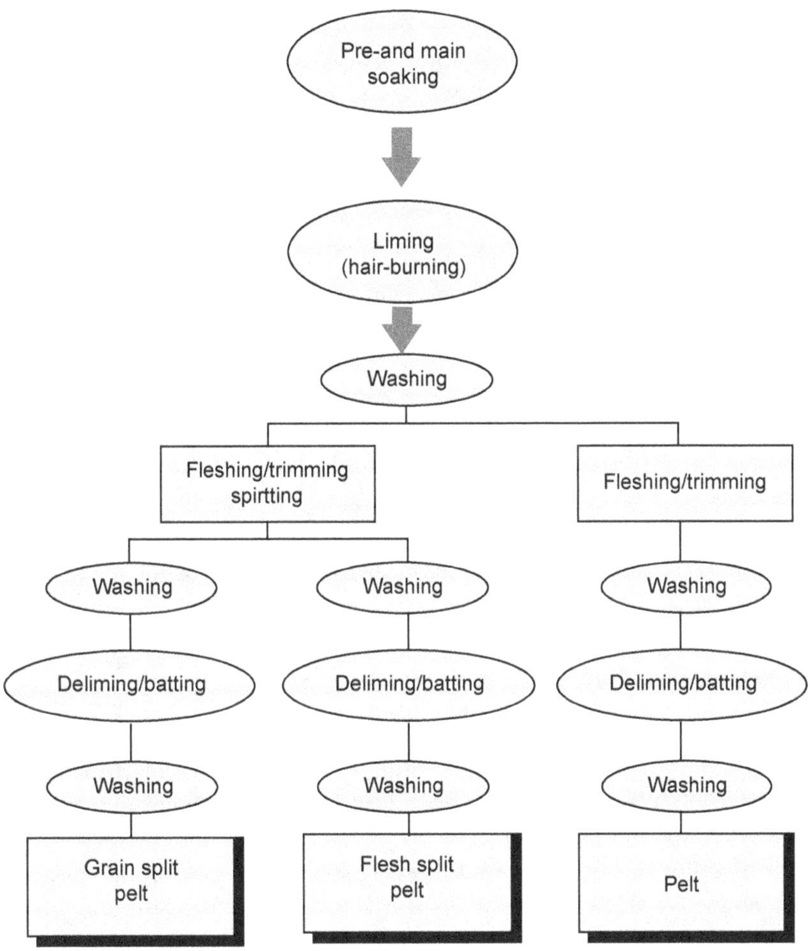

Figure 3.16 Mechanical Operations on leather

Final grading

Final grading of leather is essential before dispatching to customers. Factors like colour intensity, uniformity, feel, softness, visual appearance, thickness, design effects and natural defects such as scratches need a special consideration while grading the leather.

Measurement

Machines are deployed to measure the area of each piece of leather as the final processed leather is sold to apparel and accessory manufacturers by area.

Types of leather grades

The leather can be graded based on its distinctive grainy effect, texture, smoothness, presence or absence of blemishes, impurities and requirements of finishing treatments. Therefore, leather can be graded as top grain, full grain, split and bonded leather. The selection from available range of patent and simulated leather types is dictated by the required appearance, the style of apparels and accessories to be designed, texture and properties of the leather and the application areas.

The leather that has received some correction in any way, is referred to as top grain while full grain leather has entire grain intact without any sanding or buffing treatment. It may contain blemishes which are deliberately retained to impart a unique, distinctive appeal to apparels and accessories designed from full grain leather. The longevity and durability of full grain leather renders it more expensive and preferred choice for high-end apparels and accessories compared to its top grain counterpart. Full grain leather refers to leather which has not been sanded or buffed. Full grain and top grain leathers together are referred to as grain leather.

Full Grain or Top Grain Leather is obtained from the upper layer of a hide which undergoes splitting into layers of different thickness with the aid of splitting machine. Gauge of the splitting machine can be set during splitting to obtain layers of variable thickness. The thickness can vary depending on the gauge set at the time of splitting and the application. The outer layer with distinct natural grain if kept unattended, may show natural scars and haircell patterns. The hides after splitting are usually colored and treated with protective finishes. The full grain leather is considered to be most exotic leather type owing to its natural grainy effect as a distinct feature, strength and durability and is a preferred choice for high-end apparel and accessory designing.

Grain leathers are categorized into aniline, semi-aniline, and protected leather. Aniline leather is processed using soluble dyes and devoid of any surface pigment or coating. The natural texture and markings are retained in this leather type rendering it natural look. As far as feel and handle is concerned, aniline leather is quite light and flexible. The grain pattern in aniline leather is dependent on its source, i.e., animal species and part of the animal slaughtered for the skin.

However, absence of surface coating or pigments make aniline leather susceptible to scratching, fading, staining and soiling. Semi-aniline leather is generally treated with pigments in order to conceal blemishes, provide uniform surface, durability, making it more resistant to scratching and staining. Semi-aniline leather benefits from light surface coating due to retention of natural appearance and enhanced durability compared to aniline leather.

Protected or pigmented leather is subjected to polymer surface coating containing pigments making it resistant to scuffing or fading and most durable among all other leather types. However, the natural appearance of the leather is compromised as a consequence of pigment application. The durability of pigmented leather makes it an ideal choice for furniture upholstery, car upholstery and accessories that are intended for rough and tough use like laptop bags, sneakers, belts, gloves, etc. However, one probable demerit of protected leather is its plastic like, not so pleasing feel as a result of pigment coating on its surface.

The pigmented leather in which grainy effect is preserved before pigment coating is classified as full grain pigmented leather.

Hides which have an excessive amount of scratches or scars are further processed by correcting the natural grain of a full grain skin. The natural grain is buffed or sanded and rendered a corrected, imitated and more uniform surface texture through embossing to simulate various hair cell patterns. Embossing is a process whereby an embossed roller engraved with patterns or motifs imitating the grainy effect or texture of exotic, grain leathers is used to heat press the engraved pattern generally on pigmented leather but sometimes even on aniline and semi-aniline leather as per end-use requirements.

Consequently, the embossed leathers are flawless with no visible natural scars however, these can no longer be categorized as real or natural grain. Synthetic grain and finish reduces the original feel and durability of natural skin. An example of corrected, embossed leather is corrected grain pigmented leather which is made free of imperfections by abrading the grainy surface before the application of surface coating. The surface is then embossed with a decorative grain pattern.

Antique grain also known as rub-off leather is achieved by uneven application or partial rubbing off a contrasting top-coat revealing a lighter underlying colour thereby creating a special surface effect. The treated leather mimic the unique distressed, worn appearance of traditional leathers.

Pull-up leather also known as waxy or oily pull-up leather is characterized by a unique worn-in effect as a result of stretching during wear and hence exhibits lightening of its natural color. Nubuck is aniline dyed leather displaying tight fibre structure in the grain layer and visible grain pattern. Light abrasion on its grainy surface imparts velvety finish or fine nap while retaining the inherent grainy effect of nubuck. Sanding or buffing is accomplished to remove surface imperfections from the leather, however nubuck receives very light buffing during its processing to achieve the characteristic fine nappy effect.

Suede

The bottom layer of leather is referred to as genuine, manmade faux, corrected, embossed, coated, suede, napa, painted leather, and most commonly split leather. The split leather is obtained by splitting off the bottom layer from the grain at the grain/corium junction. The term genuine leather is in actual a misnomer for split leather as this layer is basically imitated or simulated to have texture and appearance of exotic varieties of leather such as top grain and full grain leather which are most sought after choices for exclusive assortment of accessories unlike the split leather.

Split leather undergoes slicing to make it thinner and suitable for varied applications. The former lacks the strength and durability of grain leather. However, split leather (finished split) is often polymer coated and embossed to mimic a grain leather although it is still weaker than latter. The finished split can be distinguished from grain leather by close examination of fibre packing in the structure. The fibres are much more tightly packed near the grain surface in grain leather while the fibres are loosely packed in a finished split. The low strength of finished split leather restricts its usage for accessories which are used in low stress conditions and meant for delicate handling like small pouches, wallets, hair accessories and fashion jewelry where split's low strength does not impair the functionality of these accessories.

Another variant of split leather, suede is imparted a napped finish by texturizing process. Suede offers softness and pliability which makes it suitable for designing handbags, gloves, footwear, belts, etc. Nubuck, another leather variety also exhibits nap finish and at first instance it becomes difficult to distinguish nubuck and suede. However, nubuck is grain leather which is subjected to texturizing and is much stronger and durable compared to suede.

Bonded, reconstituted, particle board leather or blended leather is obtained by shredding leather scraps and reconstituted with a filler and backed with an embossed polyurethane coating. Leather remnants and scraps from garment and footwear manufacturing units are recycled to obtain this leather variety in which fibre particles are bonded with adhesive into a fabric. Fibre bonding is thus followed by the application of a synthetic grain, hair-cell pattern and finish. Bonded leather displays suede back and imitates the smooth full grain leather. It is considered to be the lowest grade of leather, much cheaper and usually used for low-end accessories.

Split or bonded leather may be surface treated, coated with pigments, painted or embossed to closely resemble grain leather. However, accessory designers, retailers and consumers should be aware of how to distinguish between imitated and grain leather as the two vary considerably in their properties and costs. A leather that showcases a very uniform, monotonous pattern may be embossed. Furthermore, painted or polyurethane leathers are non-receptive to leather conditioners due to their non-porous structures. Imitated leathers also lacks pullup effect, which is a distinct color variation characteristic of grain leather when bent or folded.

High-performance leather also referred to as lezanova is produced by a special tanning process whereby chemicals permeate the leather to render it washable, stain resistant and water proof. Apparels and accessories designed from Lezanova offer softness and breathability.

Eco leather/ environmentally preferred leather

The environmental hazards associated with leather manufacturing can not be undermined which calls for stringent actions on the part of leather manufacturers, tanners, leather apparel and accessory manufacturers to comply with eco-friendly norms so as to minimize any environmental issues. Eco-ness of leather is gauged by the absence of certain restricted chemicals like banned azo dyes, PCP, chrome VI, formaldehyde, etc. Therefore, eco labels, high street retailers or retailers who want to have cutting edge over their competitors through product positioning have been encouraging adoption of sustainable practices during leather manufacturing and processing.

The key elements determining an eco leather/ environmentally preferred leather includes

- Control of manufacturing method of leather & inputs required for its manufacturing
- Clean technology & non-toxic chemical selection during processing
- Effective management of restricted substances
- A measure of the end of life impact

Cleaner production applications include green hide and skin processing (supply of raw material from slaughterhouses without preservation, e.g., salting), water management (use minimum volume of process water), recycling (e.g., in liming) and chromium recovery (after tanning), hair saving (to reduce dissolved solids in effluent) and application of environmentally friendly chemicals (e.g., enzymes). Special attention is also given to occupational health and safety (OHS) in tanneries. The practices followed by tanners coupled with chemical selection can go a long way in overcoming the adverse impacts paused to environment during leather manufacturing. Accordingly, following aspects need special consideration for hazzle free leather processing and manufacturing:

- Energy consumption
- Air emissions
- Waste management (hazardous and non-hazardous)
- Environmental management systems
- Water consumption
- Control of manufacturing processes
- Effluent treatment
- Chrome management
- Traceability of material

While selection of inputs for the manufacturing process, usage of materials that could improve eco profile to leather should be recommended. Elements which can be controlled at input stage are as follows:

- Usage of biodegradable wetting agents for soaking
- Reduction in sulphide processing
- Non-synthetic or polymeric re-tannage systems
- Preference for natural dyestuffs
- Vegetable oil based fatliquors
- Ability of Biodegradation within year's time span

3.2.2 Fur

Fur utilized for apparel and accessory designing, is a natural material obtained from fur bearing animals having hair, fleece, or fur fibres attached to their hides or skin. However, fur can be classified as natural material only when it is obtained from fur bearing animals. The skin along with fur when treated with toxic chemicals to convert the petrifaction prone raw skin into a usable, durable and wearable material is no more considered natural fur.

Fur-bearing animals are trapped or raised during winters for their fur. Skins converted into leather or those that have hair, fleece or fur fibres removed do not belong to fur family. Fur is composed of coarse, long and lustrous outer guard hairs and an underneath layer of underfur. Underfur is soft, downy, and provides insulation and warmth properties. Pelt is the animal hide or skin with the fur fibres attached to it. Minks and foxes are the main sources of pelt.

The classification of furs are based on two parameters:

animal families which serve as fur source - weasel family- mink, fox, bear, dog, cat, rodent, hoofed and salient fur characteristics like hair length, warmth, durability and rarity that make it stand out of any imitated versions. Durability refers to the degree of continuous shedding or hair loss of these pelts once fur is transformed into apparel or accessory. Furs vary in their ability to withstand shedding and accordingly, classified as durable and perishable.

Mink

Mink belong to weasel family. Female mink offers smaller, softer and lighter pelts compared to their male counterparts. Consequently, female skins are more prominently used for footwear, scarves, capes, kaftaans and beanies (a hat style worn in winters) compared to male mink skins. It provides dual advantages of being light weight as well as warmth.

Mink is a very durable fur with shelf life of more than two decades depending on the quality and care taken to prevent its unhairing. The most exotic furs are the one that are utilized in their natural form without dyeing. Dyed mink represents lesser quality skins used for low priced apparels and accessories.

Natural ranch mink possesses naturally lustrous sheen and consists of silky and even guard hairs . The underfur beneath guard hair is dender, compact and paler in color compared to latter. The natural rank mink is high-priced thus mutation mink is a preferred choice where economic constraints need consideration.

Mutation mink the price of mutation mink depends on its easy availability or rarity, color and uniformity of hairs. Generally, the mutation mink is composed of silky and even guard hairs with clear and uniform color.

***Pieced mink**:* pieced mink is another variant of natural mink that is less expensive and extensively used for a variety of apparels and accessories. The mink when used for textile applications is reinforced on the leather side with nylon or ribbon at points of wear.

Fur, like leather also requires treatment for its long time preservation and making it suitable for a range of apparels and accessories. Fur preservation and dyeing is accomplished by use of a variety of salts along with ammonia, formaldehyde, hydrogen peroxide, and other chromates and bleaching agents.

Fur has been considered a luxury since time immemorial owing to unique characteristics it features such as fur density, color, size, etc.

The pelt demands processing before it can be used for textile applications. Pelts are subjected to dressing process which is equivalent of tanning process, softening and preserving the hide keeping the fur intact. Dressing process includes dyeing, shearing and texturing.

Transformation of raw pelt into wearable furs involve highly specialized steps depending upon the size and type of the accessory or garment. The exceptionally high insulation and warmth properties of fur makes it a preferred choice for winter wear knitted accessories. Winter scarves, socks, stockings, gloves, mittens, etc. designed from angora rabbit hair are hot selling products among fashion conscious consumers.

A ranched fur characterized by special colors and quality is obtained from fur bearing animals bred in captivity.

The processing of fur pelts impart them softness and pliability making them suitable for myriad of textile applications. Once processed, the furs are ready to be crafted into beautiful apparel or accessories.

Fur is fragile and must be handled with care. Unlike the tanning of leather, which involves removal of hair completely from the hide, fur dressing and dyeing is intended to protect and preserve the fur hairs and follicles.

Preliminary step involved in processing fur is tanning the leather side of pelt to render softness, pliability and longevity without deterioration. This is referred to as dressing of the fur pelt.

Fleshing involves subjecting fur pelts to revolving blades to scrape away fat and excess leather reducing the weight of fur pelt. Thereafter, the fleshed pelts are soaked in a tanning solution. Protection of hair follicles and preservation of fur pelt is ensured by the use of gentle tanning solutions. The treatment of fur pelts with tanning solutions is referred to as fur dressing.

Dressing of fur pelts involve its treatment with table salt, water, alum salts, soda ash, sawdust, cornstarch, lanolin and other natural ingredients. Harmful effluents are avoided during the dressing process however small quantities of formaldehyde is used during dressing or dyeing to protect fur follicles. Moreover, the activation of tanning process requires application of gentle acids like acetic acid on the fur pelts. Release of waste water from tanning vats is preceded by skimming of fats and neutralization of pH levels as a measure to control any environmental hazards.

Pelts are further softened by addition of natural oils (vegetable oil, lanolin) to the leather side, after their removal from tanning vats. The pelts are then subjected to drumming where softness is imparted to pelts and any extra grease is removed by placing into large revolving drums, with hardwood sawdust.

The dressed fur pelt that is soft, pliable with its natural beauty enhanced can be used in this form for apparel and accessory manufacturing. Else, furs can undergo plucking, shearing, dyeing, or any other processing treatment for value addition and incorporating suitability for a specific textile application.

The unprocessed, true fur pelts are composed of two main hair types. The part of the fur visible to naked eyes is guard hair which is long, relatively course fur that protects animals from natural elements and other hazards. A layer beneath guard hair is underfur also referred to as duvet. Under fur exhibits softness and dense structure that provides warmth and acts as insulator for animals.

Plucking involves the use of special hand-crafted techniques for removal of course guard hairs to provide softer, lighter-weight and bulky free fur for apparels and accessories, retaining the underfur's warmth property. Soft, lush and velvety effect is created by shearing underlying duvet into a uniform pile.

Fur dyeing

Furs are inherently available in an extraordinary range of natural colors therefore, in order to retain the natural splendor of fur, not all fur pelts would be dyed subsequently. More recently, colored furs have been gaining popularity among apparel and accessory designers and fashion buffs. Fur dyes must be gentle and ensures that no harm is inflicted to fur or hair follicles during dyeing.

New processing techniques are employed to produce lighter-weight, supple, comfortable and wearable fur that can be utilized in gamut of apparels and accessories.

Reversible apparels and accessories make use of double faced skins in which the skin side can be processed as leather or suede. Apart from traditional plucking and shearing, grooving, sculpturing and printing of fur are yet another techniques to make fur versatile for varied applications. Intarsia techniques are carried out on fur that impart a stained glass window or mosaic effect and add to the beauty of fur manifolds.

The adverse effects of finished leather products particularly those in intimate skin contact like gloves or footwear can not be undermined. Serious health hazards like skin reactions- eczema is aftermath of high levels of toxic chemicals like hexavalent chromium used during leather production process. Moreover, the exploitation of animals for their hides, fur, fleece and hair fibres is unethical and cruelty embarked on in the pursuit to fulfill fashion requirements.

The apparel and accessory designers and manufacturers are thus poised to seek sustainable alternatives for leather and fur considering the well-being of man-kind, fauna and eco system. Designers like Gucci, Versace, Ralph Lauren, Calvin Klein, Tommy Hilfiger have discontinued the consumption of fur in their apparel and accessory merchandise and have been ardent vegan followers, exploring sustainable alternatives that guarantees warmth, insulation and aesthetic appeal matching that of fur. Accordingly, many plant-based, sustainable, and renewable fabrics like hemp, recycled cotton, bamboo, and linen are increasingly being used in fashion arena. Polyurethane, ultrasuede, pineapple leaves, grape-derived leather, cork are some of the alternatives to leather. Fleece produced from recycled plastic bottles, seaweed, hemp and modal is also finding applications in winterwear apparels and accessories. More recently, designers are designing winter scarves, stoles, mittens, footwear, etc. with vegan wool knits obtained from polyester fleece, soy-based vegetable cashmere and biodegradable tencel .

Figure 3.17 Fashion accessories designed with Fur

3.2.3 Metals

Metals are considered to be the key components for design and development of accessories like fashion and costume jewelry, hair accessories, watches, mobile cases and handbag styles like clutches, etc. Metals are non-textile natural materials found underneath the earth's surface. The strength and durability of metals make them suitable for varied applications ranging from automobiles, satellites to cooking utensils. As far as application of metals in textiles are concerned, metals can be used for manufacturing notions like zippers, snap fasteners, rivets, buckles, handbag handles, etc. that enhance functionality of apparels and accessories. The various notions utilized in accessory industry will be covered in last section of the chapter. Moreover, apart from notions, metals can be used for designing various accessories like belts, handbags, watches, jewelry and hair accessories (Figure 3.18). The properties of metals which are generally exploited for a gamut of applications are enlisted below:

- Conductors of heat and electricity.
- Exhibit malleability which allows the metals to be beaten into flat sheets enabling them to be twisted, bent, or pounded it into a variety of shapes.
- Inherent shiny appearance in some metals while the same can also be imparted by polishing on metals which lack natural sheen.
- Exhibit fusibility and ductility.

The various metals that are generally used for accessory designing are discussed in the following section.

Precious metals- metals that are costly and relatively rare are known as precious metals. The precious metals are utilized for manufacturing of premium, high-priced costume jewelry.

Gold, silver, platinum and palladium are classified as precious metals. The precious metals are popular for their rarity which is the key pricing factor for these metals.

Gold is a metal of choice for exotic, exclusive costume jewelry. Gold in its purest form is soft, exhibits low strength and thus is usually mixed with stronger base metals like copper, silver to transform it into karat gold. The ratio of gold decreases and a varying colored effect is achieved as it is alloyed with copper, silver, etc. unlike pure gold which is 24 karats i.e. 24/24 parts gold. Gold attains reddish- yellow tinge as it is alloyed with copper while alloying gold with silver imparts it.

Platinum is a soft, extremely expensive metal and is generally alloyed with other cheaper metals to increase its strength. The exorbitant prices of the metal makes it a promising and preferred choice for diamond jewelry designers for designing engagement, wedding rings and other expensive ornaments. The knock offs of exclusive, high-priced diamond and platinum ring settings are designed in cheaper substitute of platinum, white gold that imitates the appearance of former.

Silver is relatively cheaper as compared to gold and platinum. The low cost and shiny white surface of silver makes it suitable for extensive utilization in jewelry particularly traditional types of anklets, bracelets, neckpieces, earrings, rings, etc. Silver metal is soft and thus alloyed with other metals like copper when it is to be used for jewelry designing.

Figure 3.18 Fashion accessories designed with precious & semi-precious metals

3.2.4 Stones

Various precious, semi-precious and birth stones are widely used for designing accessories particularly jewelry like earrings, bracelets, anklets, necklaces, hair accessories, etc. Furthermore, stones can be used superficially on accessories as decorative embellishment like stone studded on dial and strap of watch, stilettoes, thongs footwear and crown of hats, beanies, etc.

The stones vary in their composition i.e. different elements and compounds and thus vary in their chemical properties. The different stones need to be evaluated and verified because of difference in their constitution. The stones are measured in carats. The strength of stone is a crucial property for accessories like rings, bracelets that are subjected to rubbing against hard surfaces. The stronger stones are the ones that are heavier and have compact grain. The stones should further possess resistance to heat so that they undergo very low amount of expansion at elevated temperatures.

The aesthetics aspects of the stone namely its color, appearance and gloss is equally crucial for designing accessories.

Diamond, ruby, sapphire, emerald and natural pearls are classified as precious stones.

The chemical composition of diamond is carbon and is a preferred precious stone utilized for designing exclusive, expensive jewelry. Diamonds are available in a variety of colors ranging from yellow, grey, brown and black.

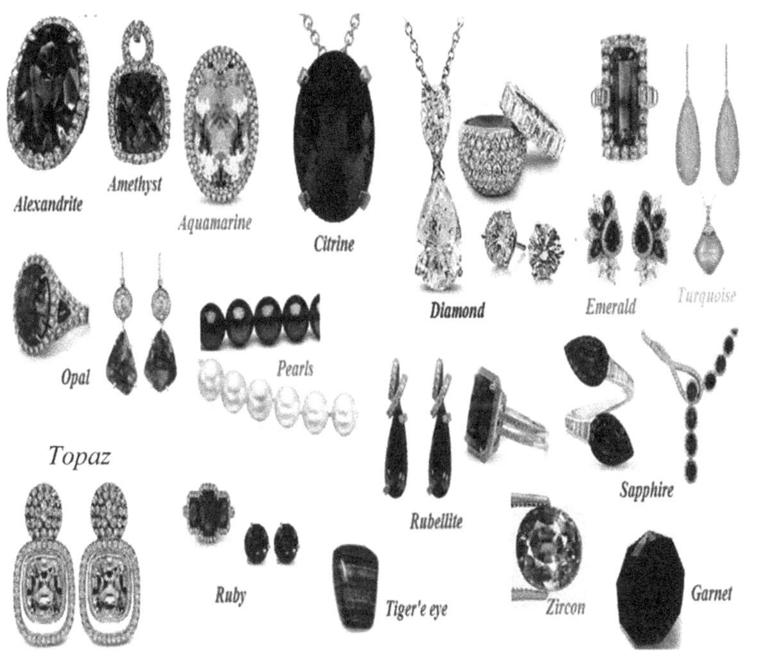

Figure 3.19 Fashion accessories designed with precious & semi-precious gemstones

Ruby is one of the valued, rare and expensive gem stone. The stone belongs to the corundum and is characterized by its red color. Oriental and spinel are the two varieties of ruby generally used for jewelry designing.

Pearls, the natural ones are obtained from an aquatic organism, oyster. A speck of sand or dust particle act as an irritant for the oyster causing it to get rid of the outer shell containing pearl. Pearls are measured in millimeter unlike other precious stones where measurement is made in carats.

Figure 3.20 Utilization of metals & stones for accessory designing

3.3 Additional components/sundries

Buttons, toggles, studs, snap fasteners, poppers, eyelets, buckles, zippers, velcro / hook and loop, frogging, hooks and eyes, magnets, grommets, brooches are innumerous functional components used in apparels and accessories. The closures such as zippers, buttons, velcros, snap fasteners are extensively utilized for accessories like slingbags, shoulderbags, clutches, knee and ankle length boots, sneakers (men's footwear style), base ball caps. Closures/fasteners serve multiple functions of enhancing the functionality of fashion accessories. Firm grip and snug fit is provided by zippers on ankle or form-fitting knee length boots while elastic and buckles on sling back thongs (women footwear styles) not just allows size adjustments but also provide comfort and ease of wearing footwear. Likewise, zippers, velcros, rivets and magnetic closures on laptop bags, wallets, clutch, sling bags ensure security and protection to belongings.

Zipper

Zipper also referred to as zip, fly, zip fastener or clasp locker is one of the most frequently used fastening device that provides complete closure via interlocking teeth or coils. It also enables binding the edges of fabric openings.

A range of zippers are available based on type of apparel or accessory on which zipper is to be attached and the required functionality. The various types of zippers used on apparels and accessories are coil zipper, invisible zipper, reverse coil zipper, metal zipper, plastic molded zipper, open-ended zippers, two way open-ended zipper, one way close-ended zipper and magnetic zippers.

The invisible zippers are composed of nylon teeth completely concealed by the fabric tape while industrial zippers are distinguished by the large heavy-duty brass teeth. Zippers serve as secure fasteners for accessories that demand secure closure preventing loss of items. Accordingly, slingbags, shoulder bags, messenger bags, soft-sided luggage, money belts and change purses are equipped with slide fasteners to ensure full security of valuables and belongings.

Zippers used on handbags are generally continuous that can be cut and made to desired length. The continuous zippers on handbags provide ease of opening and closing of handbag and flexibility in selection of slider styles – non-locking, locking, key locking, double reversible, etc.

Figure shows the main components of zippers such as top stop, slider head, pull tab, chain, tape, bottom stop, slider, etc. Zippers are composed of three main parts i.e. tape, elements and slider. As the zipper is opened or closed, the slider assist in engaging or disengaging the elements . The teeth or elements of the zipper present on each side of a zipper meshes or engages with each other on passing through the slider. The left and right side teeth engaging together are called chain. The zipper's tape can be made of any synthetic or natural fibre with polyester, vinyl and cotton tapes generally being used for textile applications.

The following section provides description of zipper components.

The top stop of the zipper consists of two permanent stays that affix to top most end of zipper and prevent the slider from slipping off the chain.

Slider head is a device that enables easy opening and closing of zipper by its upward and downward movement on the chain.

Pull tab is the part of slider that enables the upward or downward movement of slider.

Bottom stop is used at the bottom end of the zipper as a permanent stay and prevents separation of each half of zipper.

Insertion pin enables joining of the two zipper halves on a separating zipper, frequently used on jackets. Retainer box aligns the insertion pin correctly in a separating zipper

Zipper slider & pull tab - the sliding head on the zipper like key-lock, auto-lock or reversible are referred to as zipper sliders while pull tab is the lever attached to the slider.

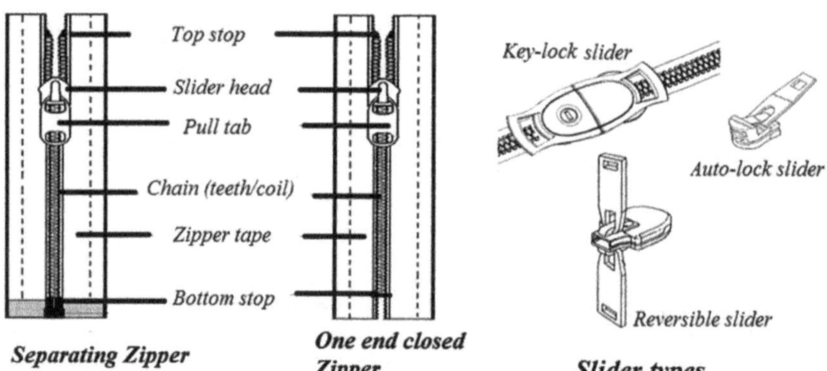

Figure 3. 21 Different components of zipper

Zippers can be classified into single-slider, double slider, non-locking, auto locking and key locking based on the types of sliders used.

A single slider zipper consists of zipper slider with only one pull.

Double-reversible slider zipper has a zipper slider with a pull tab on both sides of the slider. This zipper type finds application in reversible tote bags.

Zipper with non-locking slider has a zipper slider that is devoid of any internal locking mechanism and can slide easily without being held in its place. Purse pockets, ornamental bags, totes, boots, luggage, etc. uses a zipper with non-locking sliders.

Auto-locking slider zipper is a zipper type in which slider is equipped with an internal locking mechanism that holds the slider in place. This zipper type is generally used for handbag openings.

Zipper with key locking slider is a zipper slider which uses a removable key to lock and unlock it. The zipper finds application in travel bags, duffle bags, envelopes, portfolio folders, luggage, etc.

One-way Non-separating zipper is used for purse openings, inset zip pockets while two-way non-separating zipper also known as head-to-head zippers are used for tote bags, portfolio folders and luggage bags.

Zippers can also be classified based on tooth style, concealed (as in invisible zippers) or exposed teeth and material used for teeth construction like metal, plastic, etc. Coil, plastic and metallic teeth are frequently used for zippers. In either of the types, individual teeth are set on a cloth tape. Plastic zippers are suitable for accessories worn in marine or coastal locations like waist held pouches, wrist bands, etc.

- **Coiled zippers** also known as self-repairing zippers, have continuous spirals of nylon filament teeth set on a cloth tape. The teeth in the zipper are formed by winding of coils using coils either in spiral form with a cord running inside the coils or coils in ladder form, also referred to as ruhrmann type. Coiled zippers are usually made of polyester, nylon coil and are thus sometimes termed as polyester or nylon zippers. The zipper outshines other toothed zippers in terms of their flexibility and smoothness.
- **Reverse coil zippers** are another variants of coiled zipper which contains coil on the reverse (back) side of the zipper and the slider on the flat side of the zipper. Reverse coil zipper shows stitching on the zipper's front side. The zipper type finds application in water resistant apparels and accessories like rainwear, head covers, gloves, laptop bags, etc.
- **Metallic zippers** are composed of molded metallic teeth set on the zipper tape at regular intervals. Brass, aluminum and nickel converted into pre-formed flat wires are utilized for the construction of metallic zippers. These fasteners are generally used for apparels and accessories that demand high strength and ability to withstand rough and tough handling and washing like denims, workwear, bagpacks, laptop bags, handbags, footwear, etc.
- **Plastic-molded zippers** are similar to metallic zippers varying only in the material used for teeth construction. Plastic zippers uses plastic teeth that can be made in any color of plastic unlike the metallic

ones where the metallic teeth require painting to match the tape and surrounding fabric. A variety of thermoplastic polymeric materials such as polyacetal resin, polyethylene are used for producing plastic zippers.

- **Invisible zippers** are composed of coiled teeth construction that are concealed behind a tape making the zipper invisible on garment sections. Invisible zipper renders the appearance of a regular seamline to the attached garment portion. The color of tape and slider usually matches that of apparel or accessory on which the zipper is attached. Invisible zipper finds application in skirts, dresses, military and emergency wear providing dual benefits of appearance of a button down shirt and an efficient and easy fastening system.
- **Open-ended zippers** can have either plastic, metallic or coiled teeth and are locked on both sides via box and pin mechanism. The zippers are usually used in jackets, winter wear apparels, handbags, etc.
- **Closed-ended zippers** are closed at both ends and are often used in luggage.

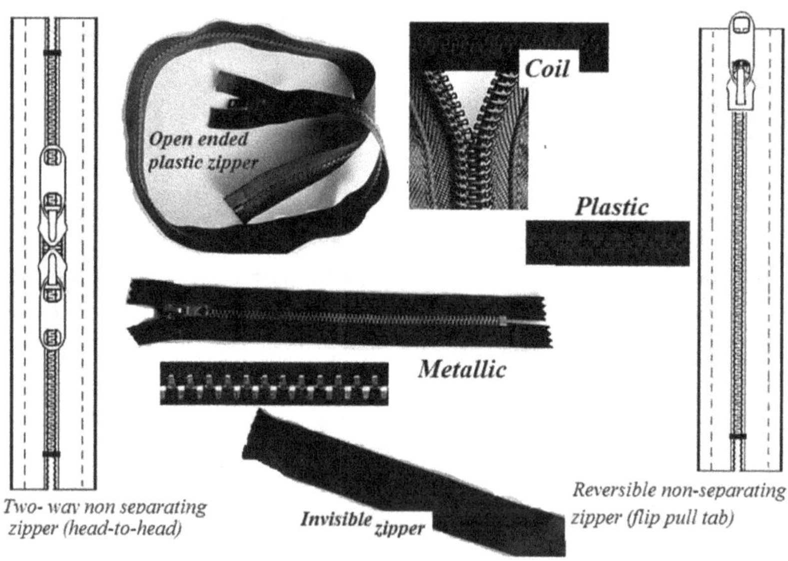

Figure 3.22 Different zipper types

Buttons

Buttons apart from being functional can also be used as decorative embellishment on apparels and accessories. Button is generally a knob, disk, or ball used as a closure or fastener. A variety of materials like antler, bone, horn, ivory, shell, vegetable ivory, wood, celluloid, glass, metal, bakelite, hard plastic, glass, stone, metal, leather, denim and canvas are used for button construction. Buttons can be attached to apparels and accessories either by hand or machine stitching. The buttons can be slipped through a fabric, thread loop, or a reinforced slit known as buttonhole. Alternatively, metallic buttons like snap, rivets ensure secure attachment by piercing through holes cut in fabric or leather for button insertion.

Bangle bracelets, clutch, crocheted purses, thongs, **stilettoes** (female footwear styles) are some of the accessories that utilize buttons either as functional or aesthetic entities.

Types of buttons

Buttons can be classified into different types based on their functionality, components, construction materials and methods of attachment to textiles. The button types generally finding application in apparel and accessory industry include shank buttons, sew through, stud and shaft buttons. The following section discusses the different types of buttons in details:

Shank buttons also referred to as self-shank buttons are composed of carved or moulded hollow protrusion on the back (shank) which provides passage for the sewing thread and enables attachment of button to the textile surface. The shank of a shank button is similar to stem of a flower holding button like flower above it.

Cloth Buttons are another type of shank buttons that can be flat or rounded with their top surface covered with fabric.

Shaft buttons are usually attached to enhance aesthetic appeal of apparels and accessories and have minimal functional significance associated with them. The shaft buttons can either be covered by a fabric or leather over a ring base for former and form base for latter.

Chinese ball button is an elaborate knot made of cord or braid.

Flat or sew-through button has holes on its top surface that enables sewing thread to pass and stitch the button onto the required substrate. These buttons can be classified into two hole and four hole buttons based on the number of holes available for thread insertion. The holes in two holed buttons are lined up parallel to the fastening edge. Four holed buttons are durable than two holed buttons and are thus suitable for heavy weight fabrics. Flat buttons can be attached either by machine or manually, however, the former method is a preferred and faster method of button attachment. Flat buttons are generally used on lightweight, casual apparels and accessories however, the buttons can be used on heavy fabrics as well by extending height of button above the fabric by forming a thread shank.

Stud buttons also known as pressure buttons, denim buttons, press studs or snap fasteners are metallic round discs, usually composed of brass. The buttons derive their name from the method of attachment on the fabric surface i.e. studding the buttons instead of using needle and thread for buttoning. The stud buttons are bipartite components i.e. couple composed of male and female studs. The fabric is sandwiched between the front (top) and rear (bottom) side of the stud buttons. The aforesaid fasteners are generally preferred for apparels and accessories constructed from heavy weight fabrics like denim jackets and trousers, denim or canvas sneaker footwear, clutch, tote bags, satchel and shoulder bags, etc. Stud fasteners provide a secure fastening compared to shank or flat buttons as the former rely on a metal rivet attached securely to the fabric. The attempt to remove stud buttons from attached substrate usually result in disintegration or tearing of fabric due to firm security with the fabric surface.

Snap fasteners first made appearance as novelty fasteners for men's trousers however in the recent times the fasteners are finding applications in a myriad of ornamental as well as functional accessories. The aforesaid fasteners are used as closures or fasteners on accessories ranging from denim handbags, belts to gloves, watches, casual hats, footwear, etc. The fastener is composed of a pair of interlocking metallic or plastic discs. A circular lip beneath one disc fits into the groove on the top disc providing a secure fastening until force is applied to unfasten or open the closure. The fasteners derived their name from snap sound as the two discs are brought in contact with each other or joined together.

Snap fasteners are inexpensive substitutes for zippers or buttons and may be concealed or revealed depending on the functionality and aesthetic appeal desirous in accessories. Snap fasteners lack the holding power of buttons or zippers, however they assist in quick opening and closing which is a must sought after requirement for kidswear, disabled or elderly people.

Toggles

Toggles are long and skinny wooden buttons with a hole for attachment on fabric, can have a rounded edge and attached through a loop instead of button hole.

Grommets

The metallic eyelets frequently used on apparels and accessories for lacing reinforcements are referred to as grommets. Grommets are tightly inserted in a hole cut through fabric, leather, canvas or any other constructional material required for designing accessories. They can be flared or collared on each side to hold them in place and are available in range of materials like metal, plastic, rubber, etc. A small sized grommet generally used in footwear for lace fastening is called an eyelet. Grommets function well in conjunction with braids or laces and together form a fastening ensemble in handbags, footwear, belts, hair accessories, jewellery and soft structured hats, like fishermen hats. Grommets aid in protection of base fabric from tearing as the laces, braids are inserted and tightened through them. Another objective of grommeting is heat and air circulation through the holes cut into fabric secured through metallic washers thereby ensuring wearer comfort.

Hook & loop fasteners (Velcro)

Hook and loop fastener also called velcro is a tape like fastener composed of two nylon tapes having raised pile surfaces. One of the tape is equipped with hooked nylon bristles while the other consists of nylon loops. The bristles on one tape meshes with the loops on the other to provide fastening as they are compressed together. The pressing of the two components together causes the hooks to catch the loops creating fastening temporarily for the time they are pressed together. On separating, velcro strips produce distinctive ripping sound as they are pulled or peeled apart.

The hooked tape of the fastener is a roll or reel of hook material that can be cut-to-length and should exhibit optimum shear and peel strengths to serve as an effective and durable fastener.

Velcros have replaced laces, braids and buckles in water tight garments (wet suits), cold protective clothing and workwear, kidswear, kids footwear, wrist bands of watches, hair accessories, base ball caps, etc. owing to the ease of fastening and unfastening unlike other fasteners. Moreover, the fasteners composed of soft hook and loop tapes reduce bulk and add comfort and can be easily hand or machine washed and thus are particularly suitable for lightweight fabrics and garments. However, accumulation of fluff or lint as the hooked tape engages on apparel or accessory portion other than its looped counterpart and thus reducing the fastening ability is the major drawback of this fastener. Therefore, more recently snag free sew-on fasteners that are one piece hook and loop assembly (with the soft loop buffering the hooks) are effective to reduce snagging and lint buildup on the fasteners.

Hook-and-eye closure is another fastening or closing element generally used for textile applications. The fastener is composed of a metallic hook made of flattened bent wire and an eye or eyelets which allows fitting of hook into it.

Another type of hook-and-eye fasteners can be tape type in which two tapes form the fastener. One of the tape is equipped with hooks while the other contains eyelets. The two tapes containing hooks and eyes can be zipped and unzipped as the tapes engage or disengage.

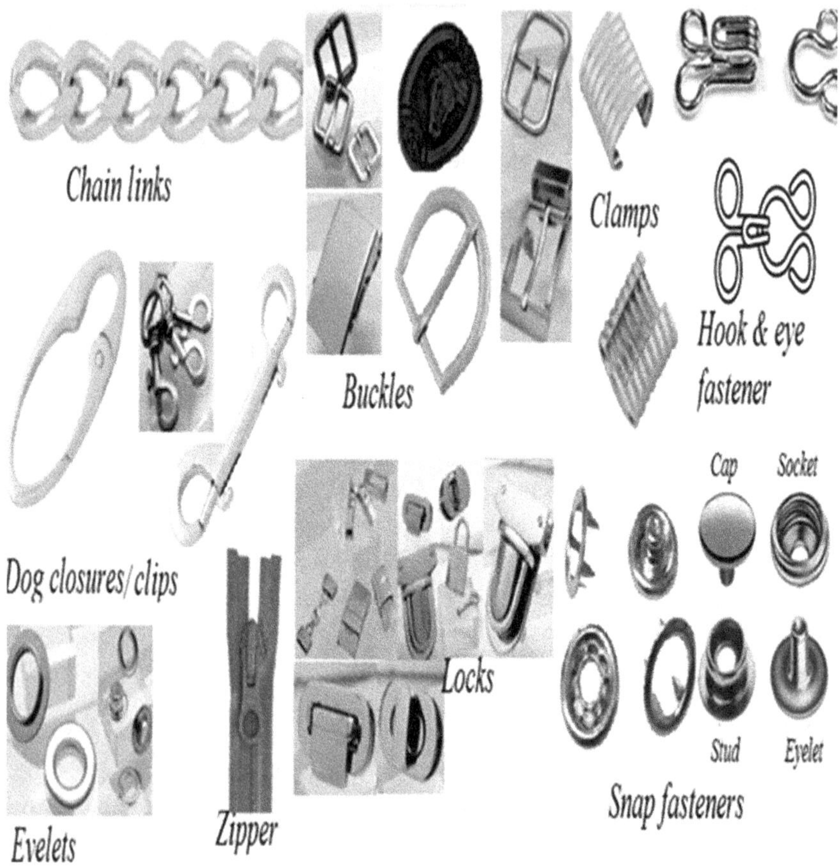

Figure 3.23 Different notions used in accessory designing

Buckles or clasps

Buckles can be used either for size adjustment or as decorative embellishment on belts, footwear, handbags, watches, hair accessories. The buckles are available in range of materials like plastic, wood, metal, glass, tortoise shell, mother of pearl and rhinestones. As far as their construction is concerned, buckles can be elaborate and ornate with carved design or loggo, gem studded or gold, silver toned or electroplated.

Frogging fasteners

Frogging fastening or frog fastener is a decorative, braided loop fastener composed of frogging loop (shaped fabric tube) and a frog (button) passing through the frogging loop. The fastener is widely used on neckline, mandarin collared upper torso garment, plackets of jackets, overcoats, etc. both as closure and decorative embellishment. Frog fasteners are also used in various handbag styles like clutches, slingbags, shoulder bags, etc.

Brooches

Brooches are widely utilized as closure on edges of neckline slits, cardigans, accessories like scarves, hair accessories, belts, etc. The brooches are composed of a catch and a pin, which are fastened to the base material (used for apparel or accessory construction) like a safety pin.

Magnetic fasteners

Magnetic fasteners are sewable magnet fasteners that are composed of strong magnet enclosed in small plastic pouches. The pouch along with magnet, concealed with facing can be sewn onto apparels and accessories like jackets, overcoats, shirts, clutch, slingbags, head bands, wrist bands, hair accessories and jewelry, etc. that demand invisible closures on top surface of fabric

Braids

Braids are narrow woven trims formed by diagonal interlacements of three strands of yarns, cords or laces. The braids are used for edge neatening and finishing on raw edges or exposed seams of straw hats, handbags to prevent edge unravelling or fraying.

Elastic

Elastic is composed of bundles of elastic fibres that can stretch upto 1.5 times to their length and then recover to their original length when relaxed. Elasics are generally used along waistline, necklines, cuffs of garments, footwear, belts, hair accessories, etc. for providing fit and stretchability.

Lining

A separate, but attached, supportive or inner fabric which conceals or covers the inside garment construction. Lining may be used fully or partially on apparels and accessories and may be completely or partially attached. Linings find applications in tailored garments, handbags, footwear, gloves and any such accessory which is in intimate skin contact for wearer's comfort and to conceal the inner seam and stich lines. Furthermore, the addition of lining provides support, increases the durability and strength of the designated apparels and accessories. The need of seam finishes on inside garment construction gets eliminated with the use of lining.

Fabrics used as lining materials on the interiors of handbags, footwear, hats, etc. should be considered for their weight, fibre content and types like partial or complete and method of attachment of lining to apparel or accessory. Linings fabrics can be woven, knitted and should possess smooth surface texture, should be durable, opaque, colorfast to perspiration and body oils and should preferably match the color of the base fabric.

Interlining

An additional fabric layer between the lining and the apparel or accessory's outer shell. The partial or complete attachment of interning on garment or accessory portion not only assists in retaining the shape but also provides reinforcement, stiffness, strength and durability on the attached area like shirt collars, cuffs, plackets, handbags, hats, etc. Insulating interlinings are a must for winter wear apparels and accessories where they can be sewn to the individual lining pieces before assembling the lining. Interlining can be attached either by stitching or fusing. However, fusing is a preferred method to achieve pucker free appearance along edges of apparels and accessories. Fusible interlinings can be composed of woven, knitted or non-woven fabrics.

The prime objective of interlining is providing shaping and support, enabling apparels and accessories to withstand static and dynamic stresses. Shape retention without affecting the textile properties of the shell is a prerequisite for interlining attachment.

Additionally, non-woven interlinings are also attached to provide bulk and serve as a backing over a small area for quilting or embroidered applique to enhance the aesthetic appeal of handbags, hats and footwear, etc. The added advantage of interlining for winter wear apparels and accessories is the ability to provide warmth and insulation properties as the filling material is used over a larger area of apparel or accessory.

Threads, lacing, ribbons & bows

Colorful threads, ribbons and bows are increasingly being used as edge neatening, finishing or surface ornamentation on stilettoes, pumps, d'orsay, espadrilles (footwear styles), jewelry, hair accessories, handbags, hats, beanies (winter wear knitted caps), stoles, scarves, mittens and gloves. Fuzzy, matte, shiny and complex yarns are the decorative yarns that are used for making embroided appliques or topstitching on accessories. Crewel threads consists of matte, colored, heavy woolen yarns generally used as embroided embellishment on accessories like pashmina shawls, stoles, handbags, hats and footwear. Colorful plastic coated, aluminized metallic threads provide durability as well as ornate, lustrous appearance when used on accessories for top stitching, applique attachment and embroidery. One of the widely used metallic thread, Lurex is used for embroidery and surface ornamentation on a range of accessories like scarves, hair accessories, footwear and handbags.

Lace

Laces are narrow woven decorative entities that are widely used in accessories either for edge neatening as brimmed hats, gloves, stoles and scarves with lacy edges or for aesthetic enhancement like laces attached on handbags, footwear, head bands, wristbands, bracelet and jewelry. Both hand made and machine made laces find application in apparel and accessory industry. Laces are categorised as needle point and bobbin or pillow lace. Needle point laces are formed by drawing elaborate patterns on parchment and thread basting on the pattern lines. Threads are connected by embroidering decorative stitches over them after removing basting stitches.

Bobbin lace is produced by stretching a paper drawn pattern over a pillow. Closely spaced pins are inserted through the drawn patterns into the pillow. Yarn wound on bobbins is interlaced around the pins, thereby resulting in open and closed patterns. The following section elaborates the different types of laces extensively used in apparel and accessory industry.

Allover or embroidered net is composed of mesh background covered with floral designs embroided all over. Chantilly Lace is characterized by fine floral designs outlined with silk threads using net as the base fabric. Black colored silk thread was mostly used as outline thread but more recently a gamut of colors have been used for lace construction. Embroidery on the lace is not carried out separately however it is worked out as the lace is being produced. The scalloped edges of fabric forming chantilly lace imparts a decorative frayed edges to the lace. Another variation of chantilly lace is galloon lace comprised of scalloping on both the edges.

Alencon or French embroidered lace is composed of net background embellished with raised motifs covered with beads and cords. Cordonnet, heavier silk cord is used to outline the lace's motif crafting a raised outer edge on a sheer net background. Alencon or French embroidered lace is chantilly lace modified to impart a raised effect through cords and beads and is a preferred yet expensive choice for bridal wear apparels and accessories namely wedding gowns, veils, gloves, etc. Venise Lace also known as Venetian Lace is a very fine lace distinct from other lace types with a floral or geometric raised design all over it. Guipure lace is devoid of any ground mesh and is not done on a mesh/ net fabric. Motifs on fabric are formed by disintegration of button hole stitches in the finishing process. Lace consists of motifs that are connected by silky thread and the fabric is available as applique or trims.

Lace Applique is single motif available as single entity rather than whole fabric and can be appliqued for value addition on a range of apparels and accessories like kidswear, women crop tops, handbags, footwear, jewellery, chokers, hair accessories, gloves, etc.

French beading laces are designed with small openings through which ribbons can be passed. A wider edging lace trim is flounce lace which is extensively used as decorative embellishment on innumerous apparels and accessories.

Insertion lace characterized by two similar and straight edges, can be sandwiched between two fabric layers or can appear on top of fabrics.

Swiss Entredeux is distinct from other laces owing to its ladder like appearance provided by cotton batiste strip with evenly spaced holes.

Crocheted Lace is a hand crafted variety of lace earlier developed to mimic the appearance of an expensive venetian lace.

Tatting lace or the shuttle lace is a delicate handmade lace formed by the process of looping and knitting of warp and weft cotton threads using a tatting shuttle . The intricate, elaborate designs created as a result of tatting finds application as edging for handkerchiefs, collars, handbags, scarves and stoles.

Tassle Lace is a type of fringe lace composed of fringed knots (tassels) hanging all along the lace's edge either at one or both sides. Tassels can be of varying sizes, colors and materials anchoring to the lace edge and rendering an ornate, appealing look to belts, shoulder and slingbags, clutches, scarves, stoles, hair accessories and fashion jewelry.

Ribbon Pass Lace is composed of regularly spaced holes for ribbon insertion through them and generally used in heirloom embroidery technique.

Beaded lace is formed by beading metallic, pearl beads on one edge giving the impression of embroided beads onto fabric.

Elastic Lace is stretchable lace primarily composed of stretchable material for its construction like spandex, lycra, elastane fibres. Fabric headbands and hair accessories, belts, wrist bands and lingerie utililize the aforesaid lace owing to its stretchability.

Smocked lace is composed of rows of smocking and can be used as decorative embellishment on yokes, waistline of garments and on scarves, stoles, hair accessories, wrist bands, belts, etc.

Gathered Lace with one edge gath**ered, resembles gathered fabric in its appearance.**

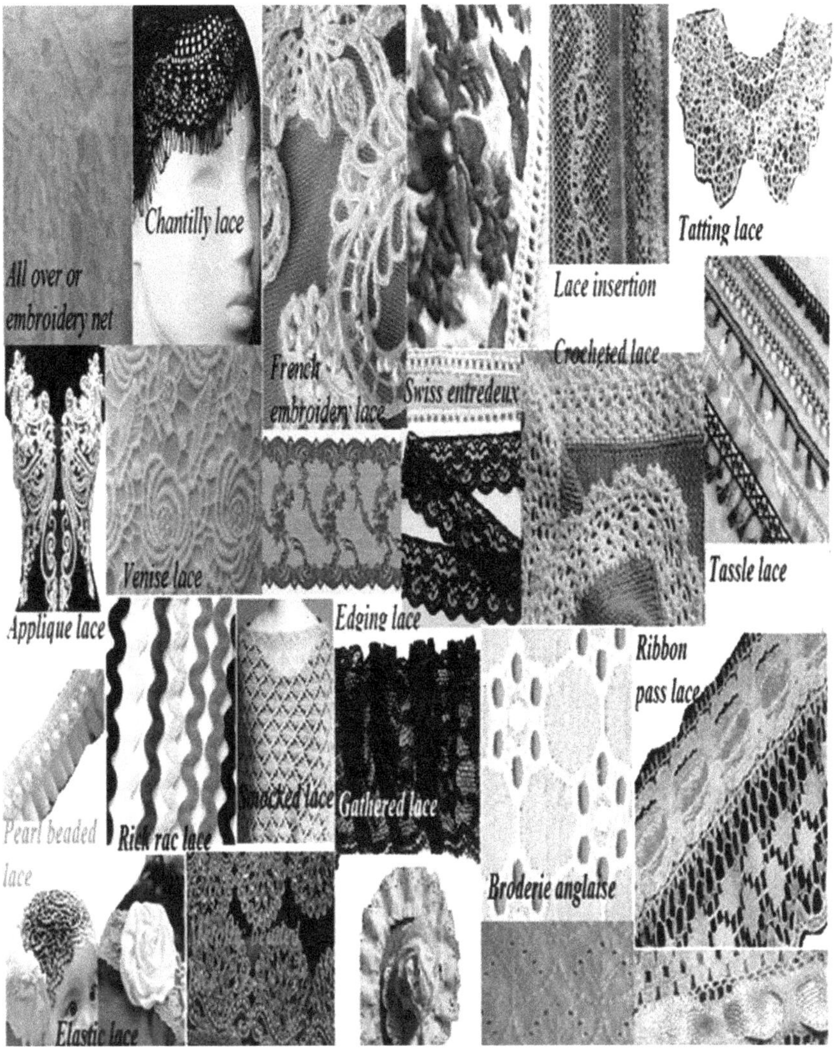

Figure 3.24 Different types of laces used for accessory designing

Ric Rac lace has a distinct zig zag pattern unlike other lace types which generally display straight or scalloped edges. Rickrack is a trimming used as a edge neatening finish on a variety of apparels and accessories to impart a zig-zag effect on the applied garment or accessory portion. Rickrack is a flat narrow, zig zag braided structure usually made of cotton or polyester. The zig-zag configuration of rick rack repeats after every third of an inch and are available in huge arrays of colors and textures and stitched or glued to the garment edges.

Cotton, silk and metallic rick rac laces are extensively utililized for edge neatening or as a seam insertion in slingbags, hair accessories, fashion jewelry, belts and other such accessories. Furthermore, this lace type is also used along necklines, sleeves, hemlines and plackets of women and kidswear. Broderie Anglaise can be considered cotton embroidered lace although it is not exactly a lace and is composed of eyelet and cutworks on white cotton, linen fabric which makes it resemble lace.

Lace like effect on fabrics can be recreated by various other techniques namely cutwork embroidery, hardanger embroidery and drawn thread work.

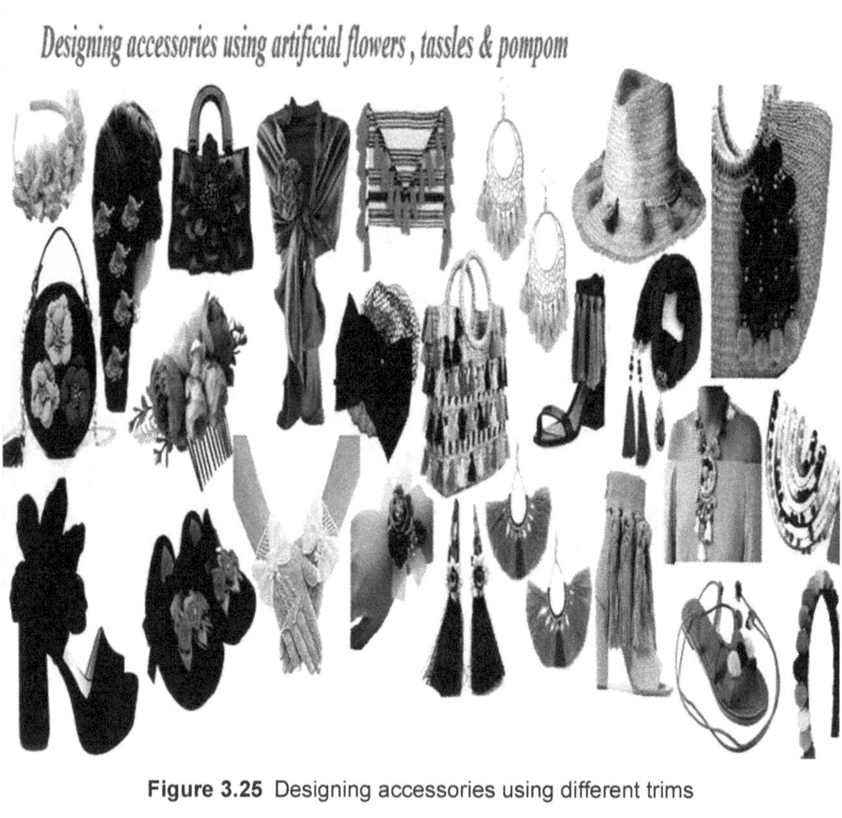

Figure 3.25 Designing accessories using different trims

4

Construction Methods & Styles of Functional Fashion Accessories

4.1 Introduction of functional accessories

Functional fashion accessories are classified as an assortment of wearable, drape able or handheld accessories which are utilitarian entities either serving as protective wear or accessories intended for a particular application like footwear, hats, gloves, socks, sportswear (athleisure). Furthermore, the functional accessories are an integral part of mankind's existence making lives simpler and comfortable (of course secondary to clothing) like watches, handbags, belts, etc. The chapter will discuss the designing, construction methods and styles/classification of aforesaid functional accessories.

Designer's prime concern before designing any functional accessory should address concerns of manufacturers, retailers and consumers. The contribution of fashion accessories to business and profit making to their creators, merchandisers and retailers is of utmost concern and no such design would ever be encouraged by manufacturers which has no takers or is just suitable for runways. Thus, a lot of emphasis is laid on market analysis, trend analysis, street fashion, forecasting trends, front and forward visual merchandising of fashion accessories during the designing and manufacturing phase.

Furthermore, a thorough examination and understanding of market segmentation is crucial for the designer to come up with elegant, trendy designs in fashion accessories. Accordingly, the market is divided into demographic and psychographic segments. Classification of consumers by objective data like age, education level, geographic region and annual income encompasses demographic segmentation. Psychographic segmentation is based on consumer classification by subjective data like lifestyle, keenness and willingness to accept new fashion, values and individual preferences. Designers, thus considering the segmentation choose the colors, materials, prints, motifs and styles for the functional accessories to be designed.

4.1.1 Introduction to footwear

Footwear is an accessory category worn on the feet both for protection against the external environmental elements, ground textures and as a means of adornment to compliment the look of wardrobe. Footwear primarily serves the purpose to ease locomotion and prevent injuries. The footwear industry is a fast growing and competitive industry that targets to capture the niche market by mass customization of footwear catering to the consumers' aspirations for style, fit and comfort. The footwear manufacturing companies thus need to act pro-actively to develop a system for rapid design changes in sync with newly introduced apparels lines, creation of myriad of virtual footwear designs and adoption of mass-customization principles to make the footwear manufacturing process cost effective.

Figure 4.1 Footwear styles using different textile raw materials

Footwear designers give precedence to functional and aesthetic aspects of the footwear while designing any footwear. Thus, the footwear manufacturing units cater to the needs of both dress and casual shoes. Consideration of attire with which pair of footwear are to be matched, the occasion on which footwear are to be worn and any desirous specific functionality like footwear intended for patients suffering from foot pain and disorders is crucial while designing footwear. Accordingly, footwear are designed in a variety of assorted range of materials ranging from patent, exotic leather to imitated leathers, suede, polyurethane, canvas, etc. The materials chosen for footwear construction are determined by the intended functionality. Figure 4.1 showcases the footwear styles using different textile raw materials.

Thongs, pumps, stilettoes, espadrilles are casual or occasional female footwear styles that uses suede, satin, sateen, jute and canvas fabric and are usually embellished with trimmings like ribbons, laces, macrame, semiprecious gemstones, etc. However, men's footwear styles like moccasin, wingtip, oxford are designed in leather, imitation leather with minimal of trimmings like grommets, braids, etc. Likewise, sneakers or sport shoes are designed considering the comfort of sportsperson by using a permeable material in upper and lining of footwear to ensure sweat dissipation easily.

4.1.2 Required properties for footwear material

The materials used for footwear construction should be strong, durable, flexible, possessing high tear and abrasion resistance and easy to care and clean during their lifecycle as the footwear in use will be subjected to significant amount of stress, abrasion against rough surface like pebbly ground, damp, muddy surfaces, etc. Moreover, the footwear should not impair the normal functioning of wearer and should be comfortable against the skin. Accordingly, construction of footwear demands such textile and non-textile materials that exhibit optimum tensile and tearing strength, abrasion resistance, water resistance and porosity enabling rapid sweat dissipation from wearer's feet. Among non-textile materials, leather has lion's share of market as far as footwear construction is concerned. Leather is a preferred material for footwear construction owing to its appealing look post processing, shape retention, mold ability and durability. Top grain, split, slab, aniline dyed and embossed leather are generally used to design footwear ranging from moccasins to loafers and oxfords. The leather processing and different leather types have been discussed in detail in Chapter 3.

4.1.3 Materials for footwear

A wide variety of textile materials like jacquard woven, brocades, canvas, denim, jute fabrics along with knitted fabrics can be used for the outer body of the footwear. The selection of particular fabric type depends on the occasion and usage of footwear.

Leather by far is the most preferred material for shoes and shoe lining. Conformity to feet's shape and breathability makes it designers' choice for footwear particularly dressy and formal styles. Calf, side kid leather sheepskin, pigskin are generally used leather types for footwear uppers and linings while snake and fish skin are exotic leather varieties. Value addition and surface ornamentation of leather can be achieved by application of natural, shiny, patent and suede finishes. Apart from the exotic look, durability and breathability of leather, consideration of stretch direction is essential as most applications demand less stretch from head direction to tail than sideways. Less stretching from toe to heel than across the shoe referred to as tight to toe is ensued by effective lasting process and shape retention

Synthetic materials like polyurethane, PVC are also gaining attention of footwear designers and manufacturers to cater to budget footwear market. The look of these synthetic materials closely resemble leather however, they lack the durability of leather.

Furthermore, textiles are increasingly being used for constructing various styles and components of footwear. An array of textile materials from woven, knitted, narrow woven fabrics have made their presence felt in footwear industry both as functional and aesthetic entities. Satin, dress fashion fabrics matching the wardrobe have been popular for designing occasional, bridal and formal footwear.

The season and wearer's physical activity commands the selection of color and fabric texture for footwear. Linen look fabrics, ribbed weave fabrics; knitted breathable fabrics in cool colors are preferred for spring/summer footwear line while wool, flannel fabrics are preferred for winter shoes. Casual, daily wear footwear can be designed with canvas, denim, jute fabrics while for athletic sportswear generally moisture absorbing, breathable knitted fabrics are preferred.

Moisture repellant fabrics are used for designing outdoor footwear intended for adventure sports like snow skiing, hiking, etc. Nylon owing to its inherent moisture resistance, strength, light weight and tight weave is preferred for outdoor footwear.

Gore-Tex, a nylon or polyester base fabric laminated with Teflon exhibits superior wicking and moisture transmission properties. The laminated fabric is designed with very small pores that resist permeation of water droplets or wind through fabric. Sport shoes, sneakers designed with Gore-Tex enables quick sweat absorption and dissipation from wearer's feet providing dry feel next to skin.

Footwear designed with lining fabric like Thinsulate repels the moisture and provide thermal insulation to wearer's feet. The fabric is a preferred choice for boot linings owing to its light weight and warmth retention properties better than down feathers.

Woven textiles composed of natural, synthetic and blended fibres are used for footwear construction. The strength, durability and dimensional stability of woven fabrics make them preferred choice for trainers, casual shoes and sports footwear. Silk, silk satin, cotton, linen, polyester, nylon fabrics are some other options while designing dressy footwear particularly peep toe, thongs styles which are adorned as evening and occasion wear, summer wear.

Backers and reinforcements used in footwear are usually made from finely woven cotton fabric with an iron-on adhesive. They are used to add thickness and control the material to prevent it from stretching.

Denim and canvas fabrics are preferred for sneakers, ballerinas or any footwear style which would follow rough and tough usage. Brocade fabrics are used to accentuate glossy, shiny look for dressy, party wear female footwear. Jacquard woven and printed fabrics backed with lining materials are relished for female and kids' footwear alike to flaunt a range of colors, prints complimenting the attire.

Knit structures owing to stretch ability, excellent moisture and air permeability and comfort properties are increasingly being used for sports shoes and sneakers where the sweat generation on wearer's feet is quite prominent. Hence, knitted fabrics used for footwear construction enable sweat absorption from wearer's feet and provides comfort to wearer.

The main advantage with textiles is their exact, known dimensions and properties like weight, thickness and area or yardage. Another advantage is the cost-effective pattern-making, cutting and spreading process during footwear construction as textiles can be spread to form a lay composed of several plies for pattern cutting. However, a single layer needs to be worked during spreading and pattern cutting when working with non-textile materials like leather making the process cost intensive. The textile materials apart from being used for construction of shoe uppers, also find application as stitching, fastening, gripping, supporting and aesthetic elements of footwear like sewing thread, velcro, zippers, elastic, narrow woven fabrics like laces and braids, linings and net flowers, tassels, applique, burlap, etc.

Non-wovens

Non-woven textiles namely felt and leather look synthetic coated materials fall in the category of non-woven textiles. Non-woven structures are composed of random fibres sealed or bonded together by application of pressure and adhesive.

Coated non-woven is composed of coated synthetic compound spread on top of non-woven materials. The materials however, lack the shape retention property of leather particularly when wet. Felt in damp condition loses its shape entirely and thus its usage is limited to indoor footwear. A structure composed of coated synthetic top layer and knitted back layer is used for lining and sock of the footwear, thus providing an economical and vegan option for the consumers.

Felt and leather look synthetic coated materials are also utilized for footwear manufacturing. Coated materials on account of their wash ability, availability in vast color ranges and lightweight are preferred for footwear.

Felt, a variety of wool/synthetic fibre based non-woven fabric is produced by matting, condensing, needling and pressing. Felts are available in varieties of softness, colors, shapes and sizes. Non-Woven polyester fabric saturated with a heat formable plastic is used for toe counter or toe puff.

Recreational or professional sports shoes should meet the requirements of functionality and quality and thus should be designed to exhibit lightweight, durable and high-performance characteristics. Performance Materials made from non-wovens are specially tailored for use in sports footwear in eyelet reinforcements, insoles owing to their ability to fulfill all the quality aspects desirable in sports shoes. The materials not only exhibit low weight but also minimizing environmental impact owing to high proportion of recycled raw materials. Micro fibre or conventional needle punched non-woven fabrics on account of their excellent abrasion resistance find application in heel grip of the footwear.

Non-woven fabrics are also used for making backers. Tea bag and swans down backer are non-woven backing materials to provide extra support, strength and thickness to fabric so that materials can withstand stresses imparted during manufacturing processes and prevents excess stretching of material. Non-Woven fabrics are used for shoe uppers, lining and heel grips ensuring high durability and softness to wearer. The non-wovens used for such application can be embossed, soft and hard finished. The fabrics are manufactured using various techniques like needle punching, chemical and thermal bonding process.

The following salient features make non-woven fabrics suitable for footwear construction:

- Exhibit colour and dimensional stability
- Do not show fraying and unravelling at edges
- Easy to slit, die-cut, sew, seam, glue, laminate and trim
- Light weight and soft texture
- Ease of use, handling and care
- Improved adhesion

- Can be easily treated with anti-static and softeners
- Possess high stretch ability, tear, and breaking, puncture strength and abrasion resistance
- Chemically inert

Performance materials

Material innovation has a path breaking role in footwear designing right from main-street fashion footwear to athleisure and performance sports shoes. Footwear worn in damp conditions like during sailing are made of wicking mesh that provide dry feel to wearer. Likewise, breathable mesh in running shoes, sneakers enables rapid sweat absorption from wearer's feet.

It is interesting to note that footwear pattern is not the most critical or the special feature, the footwear materials are! Outdoor footwear worn in cold weather conditions, snow covered regions or worn for snow sports, etc. should provide thermal insulation and warmth retention properties. Accordingly, footwear for all such applications utilizes insulating material in their construction. A state of the art footwear with inbuilt healing properties are devised for marathon runners. The footwear is lined with a silver impregnated mesh that aids in healing and stress relaxation to wearer's feet.

Backers on exotic skins, suede are made of woven cotton fabrics equipped with an iron-on adhesive. They provide thickness and prevent the material from stretching.

Performance materials namely nylon, polyurethane (PU), polyvinylchloride (PVC) are increasingly being used for footwear construction on account of their light weight, wash ability and availability in wide color ranges.

Mesh, polypropylene, neoprene, lycra, spandex are some other performance/ specialty materials extensively utilized in smart, high-end footwear.

The durability, strength, breathability and stretch ability of these materials make them suitable for footwear intended for sports, climbing, etc.

Climbing boots and running shoes make use of polypropylene rope and webbing in their laces primarily used on climbing equipment owing to their durability and indestructible properties are now being used as laces in climbing boots and running shoes.

4.1.4 Footwear Components

A number of components are used for footwear construction. The components used vary according to the styles of footwear. A mary jane female footwear style consists of strap and buckle while thongs are devoid of any buckles. Sneakers use velcro as the fastener while grommets, eyeleting and braids serve as fastener for wingtip and oxford footwear styles. The raw materials utilized for various accessory categories have been discussed in details in Chapter 3. The main components used for footwear based on their styles will be discussed in this chapter. Adhesives, buckles or clasps, counterforts, grommets or eyelets, heel, hook, insole, shank, sole, nail, treads, welt are the common components utilized for footwear construction. However, the types and number of components are style specific. Clogs or mules, a footwear style common to both males and females is devoid of any fastening element like zipper or velcro while grommeting through which braided shoelaces can be inserted, are prerequisite for oxford or wingtip styles of footwear.

Footwear comprises of uppers and lowers. The upper components aid in covering and supporting the top of foot. Upper comprises of sock lining, counter, quarter, saddle, vamp, tongue and toe cap.

Sock lining - the rough edges on the interior of shoe upper is covered by leather or nylon tricot sock lining. Sock liner is a foot bed in form of foam padded mesh on which foot stands. It is a thin piece of leather, silk, wool or EVA, placed on an insole of boot. Sock liners absorb moisture from wearers skin and keeping the footwear warmer in cold climate.

Counter - centre back seam is concealed by a decorative component known as counter. Between the lining and upper of the shoe, a stiff piece of material is placed at the heel of a shoe that helps in maintaining the shape of the shoe.

Quarter - constitutes the back portion of the shoe upper and covers the heel behind the vamp. Strengthening is provided to heel section of the quarter by inclusion of a stiffener enabling support to the foot's rear. A continuous piece of leather can be used for vamp and quarter depending on the footwear styles.

Saddle - is separate piece of upper, crossing foot over instep

Vamp - covers the feet's top in front section of shoe upper.

Lining is attached on the interiors of the footwear around the vamp and quarter in order to increase durability of footwear and provide wearer comfort. Linings are composed of kid skins, sheepskins, drills or twills. Plumping up or reinforcement of vamp and tip is ensured by cotton doublers.

Tongue - can be a separate leather piece or vamp's extension protecting the top of foot. It is sewn into the shoe's vamp extending to the main opening. Tongue provides padding, comfort and support for the feet's top. Moreover, gusseted tongues protect the ingress of grit and debris between the sock and liner.

Toe cap - serves as an additional covering over toe section of vamp

The lower of footwear consist of heel and sole.

Heel is support piece that is attached to shoe's outsole under foot's back part. Heel making process is very painstaking and complicated as perfect fitting with sole and upper is of utmost concern. Moreover, any imperfection in height or width of heel will detract the design and look of footwear.

Sole is a single layered bottom surface of shoe not inclusive of heel. Soles must be constructed from strong yet flexible material that has the ability to withstand abrasion against ground while walking and against extreme climate. Sole types can vary according to footwear styles. Thin elegant leather soles are preferred for dressy shoes while highly specialized performance soles comprise the lower of athletic shoes. Leather soles are generally handcrafted while injection molded soles are created by computerized machine equipped with temperature control.

Sole unit refers to combination of heel and the sole.

Heels - Footwear heels can be distinguished based on their height, shape and construction material. Heel height is measured in 1/8 inch increments so that one inch heel height corresponds to 8/8 heel.

A variety of materials are used for heel construction. Wood is used for low heels and thick block heels. Higher and thinner heels are usually molded from nylon or polycarbonate comprising of a central core of steel for additional strength. The exterior of heel can be covered with leather, fabric or painted. Plastic, metals, cork, Perspex, stacked leather wood, cork, rope, suede, fabric are different materials which may be used for heel construction.

The heels are characterized by following parts:

Seat- part of heel directly under foot

Breast – is the front part of heel under sole

Neck – is the back part of heel

Top lift – is the bottom of the heel that touches the ground

The heel styles based on their shapes can be classified as louis, stiletto, kitten, wedge, platform wedges and Cuban heels.

French heel also referred to as louis, curved or pompadour heel is a medium to high curved heel

Stiletto is characterized by thin and high heel

Kitten heel is characterized by a thin based, little curve, low heel. It is distinguishable from other heel styles owing to its placement and feminine shape. The heel tapers towards bottom into thin base cap.

Wedge heel is the heel style used in women's footwear features a triangular or wedge like piece extending from front or middle to back of sole. Wedge heel is characterized by low to high heels.

Platform wedge heel enable whole of the foot to be elevated by an additional 1 to 2 inches. The heel is higher than wedge heel and is a combination of platform and wedge heel.

Cuban heel – Men and women cowboy boots usually feature this heel style. Cuban heel has slightly tapered back and straight front.

Stacked heels - Glued and sanded layers of leather can be used for constructing men footwear's heels known as stacked heel.

Toe puffs & heel stiffeners Toe and heel part of the shoe require harder area around them. Toe puffs and heel stiffeners are thus inserted between upper and the lining for providing hardness at the toe and around the heel part of the shoe. Moreover, the components are added to footwear for shape retention and support for feet.

Thermoplastic materials capable of softening on heating are used to construct toe puffs and heel stiffness. The material on cooling solidifies and hardens to attain the mold shape of the last.

High-priced footwear uses preformed toe puffs and heel stiffeners made of adhesive coated fiberboard which can retain the shapes longer compared to thermoplastic ones.

Sole unit is the unit constituted of sole and heel created by molding and attachment to bottom of footwear. End-use of shoe determines the selection of materials for sole unit. Rubber cup soles composed of one-piece rubber are used as sole units for sportswear for enhanced wearer comfort when worn during specific sports. The materials and construction of sole units can vary according to footwear style and their usage.

Performance sports shoes worn for specific sports demand strength, shock absorption, comfortable wear and cushioning effect to wearer's feet. Accordingly, sole units composed of multilayer or several pieces fitted together are preferred for sports shoes.

Central layer, midsole composed of ethyl vinyl acetate (EVA) when used with rubber cup soles, provide cushioning effect to wearer's feet. Performance sports shoes like basketball footwear can be designed with several pieces like rubber outsole, midsole and heel insert fitted together.

Each piece is designed for enhanced functionality of footwear. Thin rubber outsole layer provides traction and better grip, midsole of EVA provides flexibility while heel insert of EVA is used for shock absorption.

Casual footwear is also designed considering the comfort aspects of prospect consumers and thus designers take inspiration from sports footwear styles. Sole units for casual footwear are produced from synthetic materials like thermoplastic resins (TPR) and polyurethane (PU). Lightweight sole with built in cushioning can also be achieved by air injecting of blown polyurethane during molding process. Fashion footwear like ballerinas, espadrilles, clogs use lightly cushioned sole units composed of synthetic rubber, wood, cork and rope units respectively for comfortable, bulk free footwear.

Bespoke footwear uses leather soles. Vegetable tanned leather owing to its thickness and sturdy structure withstanding wear and tear is preferred for sole of footwear.

Apart from leather, rubber – natural crepe and molded rubber can also be used as soling material. Rubber sole creates a waterproof seal by directly molding it to shoe. A range of colors and textures are available in soles composed of synthetic rubber and polymeric materials.

Enhanced comfort to wearer's feet is ensured by using a soft, spongy sole that is characterized a microcellular layer of tiny air bubbles.

Outsole is the exposed part of the sole directly in contact with the ground and thus should exhibit grip, durability, water resistance and provide protection to wearer's feet against external elements.

Insole considered to be foundation of whole footwear can be made from range of materials like leather, fiberboard. Nowadays, Insoles are composed of fiberboard. Fiberboard is composed of cellulosic fibres and synthetic rubber bonded together. The material is available in varying thickness and strength. A single flexible or blended layer of fiberboard can be used depending on footwear styles and heel height.

Two part fiberboard with softer, flexible material in front and harder material at back is used for insole of footwear featuring heels higher than ¾ inches. Strength to back portion in high-heeled footwear is provided by a thicker and denser material.

However, whole sole is made of single material for women's low heels, men's and children footwear which do not require as much strength as with high-heeled footwear.

Accurate molding of insole to curve of bottom of last is essential. Shank that serves as a bridge between back of the shoe and joint provides strength to footwear. Weight of body need to be supported by this part of the shoe and hence shank is considered a crucial component. Groove cut into the insole allows for the shank to be fitted perfectly into the curve of the insole. Low-heeled shoes use wooden or fiber board shank that provides strength while fluted steel shank are preferred for high-heel footwear that require additional strength and rigidity without adding too much bulk and thickness.

Welt - a leather, rubber, or plastic strip stitched to the footwear's upper and insole is referred to as welt. Cork or some other filling material is used to fill the space enclosed by the welt before cementing and stitching the welt to outsole.

Reinforcements applied on footwear to prevent tearing or stretching resembles in structure to non-woven backers however, the former are stronger than backers. Reinforcement tapes are available in different widths like back seam tape, topline tape.

Back seam tape creates an even surface by covering and strengthening the seams applied on footwear. Tope line tape is a narrow tape that strengthens the tope line of ballerinas, pumps and prevents any stretching of material.

Shank, a narrow reinforcement is considered to be structural element of footwear. Shank along with insole forms the footwear's foundation that strengthens the shoe between the back and the joint. Flat and low-heeled shoes use a simple wooden or fiberboards shank while fluted steel shank is used to provide rigidity and strength in high-heeled footwear.

A variety of materials namely steel, plastic, fiberglass, Kevlar is used for making shanks to be used in work boots. The shape, stiffness to outsole and a more stable fit is credited to the presence of shanks in footwear.

Hardware

Functional and aesthetic aspects of footwear are taken care of by use of variety of hardware and trimmings. Hardware used for footwear must be strong and durable. Trims although are decorative embellishments and non-functional but they must be durable and long lasting. Eyelets, ski hooks, D-rings, shoelaces, zippers, elastic, buckles, velcros are the functional components generally used on footwear based on style and functionality of footwear. A variety of trims available in several materials, sizes and shapes like metal, decorative buttons, buckles, bows, flowers, feathers and chains are generally used for surface ornamentation on footwear.

Eyelets are the metallic reinforcements that cover the holes in the facings of footwear and thus prevent the stretching and tearing of holes. Canvas sneakers, footwear with laces as gripping elements generally use metallic or painted eyelets.

Ski-hooks, D-rings used on ski-boots enable the holding of shoelaces in position.

The basic requirement for shoelaces is their strength and ability to withstand pressure exerted during tying of laces or during walking. Shoelaces are available in wide range of dimensions, materials and shapes varying from flat, oval to round.

Cotton, linen, synthetic and blended fibres can be used for shoelaces.

Zippers when used as closures on ankle, knee length or thigh length boots enable convenience of doffing and donning the boots without much fatigue to the wearer as the boots can be opened up along the whole length of leg and down the feet.

Elastics owing to their stretch ability are used in gussets providing space for footwear to stretch.

Buckles also referred to as clasp are metallic or polyamide fasteners used along with strap to provide firm, secure fastening and grip on footwear like mary jane, ballerinas, thongs, etc. The buckles should be strong and durable with an ability to withstand the pressure exerted during walking.

Velcro, a hook and loop fastener, composed of two polyamide tapes is preferred choice for elders and kids footwear owing to the ease of fastening and unfastening compared to buckles, laces, etc.

Procurement of raw materials and components

The components used for footwear vary widely depending on footwear styles. The various components used for footwear have been discussed in the previous section. Athletic, dressy, casual shoes will have drastically different components in contrast to components used for pumps, stilettoes or other occasional footwear.

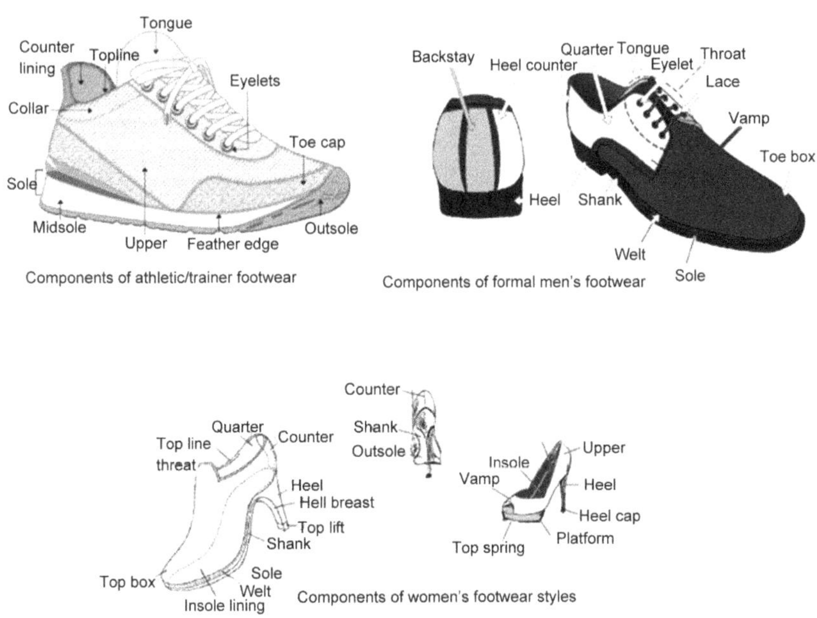

Figure 4.2 Footwear components

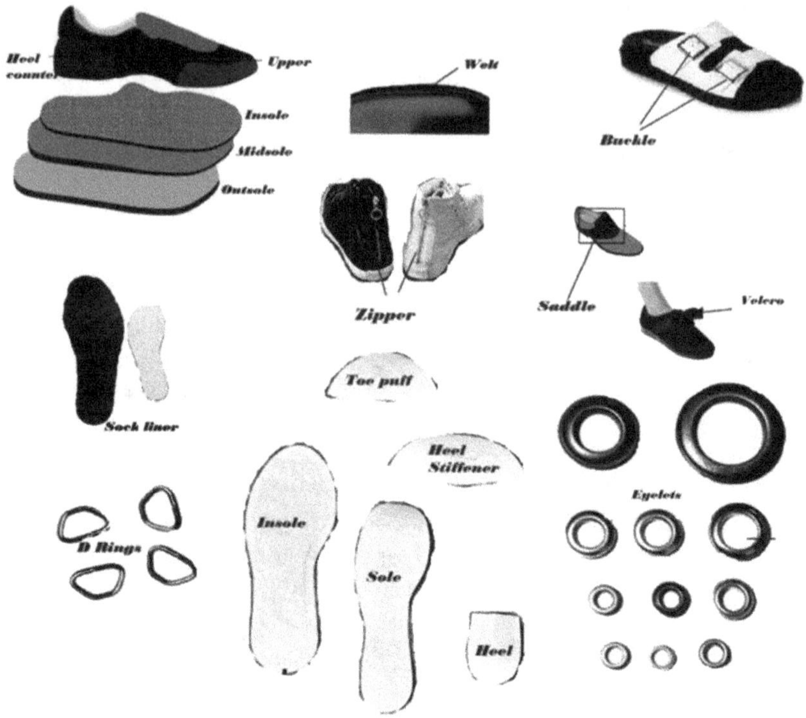

Figure 4.3 Textile & non-textile footwear components

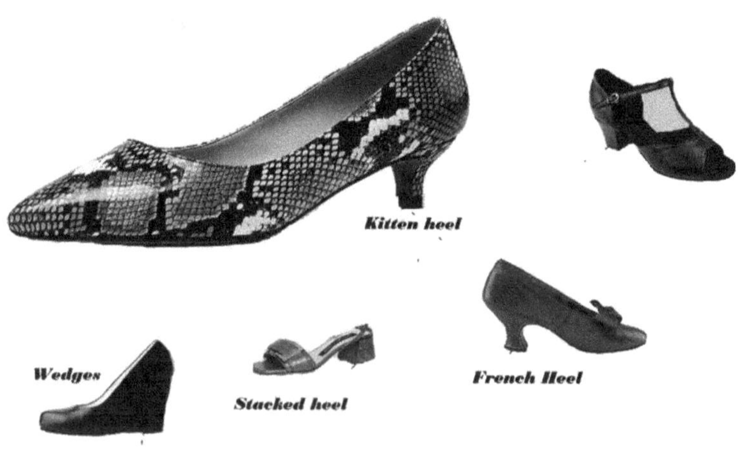

Figure 4.4 Different heel styles

Realistic timelines can only be established if designers and merchandisers have insight of sources, vendors or markets to look for timely procurement of all the raw materials and components for rapid prototyping as well as for mass production of footwear. Special arrangements need to be made for components like outsoles, insoles, leather boards, toe puffs, fillers, stiffeners, shanks, heels which may have to be procured from different manufacturing units.

4.1.5 Footwear construction

A number of footwear construction methods are there from which a shoe designer can choose. A myriad of stages are involved in footwear development from the designer's concept to the creation of finished, wearable footwear prototype. The process involves development of designer last, translation of designer last into production/development last, pattern-making, die manufacturing, pattern-cutting, fitting or stitching, lasting, assembling the remaining components, bottoming and finishing (Figure 4.5).

Designing of footwear is not possible in isolation without market analysis for trends of apparels in a particular season. Designers fetch inspiration from innumerous sources like market survey, trips, historic collections, existing fashion trends and trends in apparel line. Accordingly, the designers work in close association with merchandisers, retailers and apparel production units to be updated about the latest trends in apparels so as to design and develop footwear that coordinate well with apparels. A bridal attire designed with vibrant colors and lavish embellishment demands a pair of footwear that compliments and completes the bride's look. Accordingly, footwear designer may design high heel pumps or stilettoes in shimmery, lustrous silk, satin fabric for the upper and the look completed with semiprecious stones, ribbons, bows or flowers in colors that matches well with the bridal's attire and jewelry. Likewise, athletic shoes like sneakers are designed considering men's and women's athleisure, track pants or joggers so that footwear can enhance the sporty look.

Depending on the function and style of the footwear design, there are many options for connecting the upper to the outsole unit. Designers should thus take a note of materials and sundries (trims and notions) required as per the functionality and aesthetics of footwear along with color, shape and proportion of footwears. An important consideration is the view of designed footwear from various angles.

Comfort, ease of wearing and taking off, sweat absorption and dissipation from lining and insole of footwear, utility and aesthetic aspects are some of the considerations which are fundamental for footwear designers and manufacturers. However, textile requirements, construction problems and production economics confines and restricts the designers' creativity looking for the plausible design options. A system is thus required that enables changes in shoe-last design, exploration and mass-customization of footwear in cost effective mode. Mass manufacturing of footwear provides ready-made shoes sold at relatively affordable prices. The sizing and grading system during manufacturing process enables creation of footwear of different sizes. However, a customer with abnormal foot shape or those suffering from some foot disorder require footwear customized as per their foot type. Moreover, footwear manufacturers experience challenge owing to wide variations among population, gender and age.

Accordingly, designers need to be watchful of utility and comfort while coming up with new creations. Hence, the collaborative efforts of designers, fashion merchandisers and trend forecasters along with production engineers help in finalizing the styles that will be made up in model sizes. Communication with clothing design team enables the designers to finalize themes, colors and fabrication so that the footwear resonates with the current clothing line.

The designing for footwear usually commences with handmade sketch by designer including all meticulous details pertaining to style, components and trimmings on the footwear. The sketches are followed by development of mood boards for individual concepts by designers which are then used for concept presentation to merchandising team, sample makers and manufacturers.

Hand sketch, drawing on the last or computer aided sketching techniques are used by designers in design process before getting the prototype approval. The handmade sketches have however, been supplanted by designing using computer- aided designing systems. CAD systems is advantageous owing to the capability of the system to produce both two-dimensional upper design and three-dimensional last (a wooden or polyethylene replica of footwear), projection of drawing on the last and grading of footwear patterns to achieve varying sizes of footwear.

Designers further need to have follow ups on lab dips, wash finishes and stitch quality to ensure that prototype and final footwear conforms to designer's initial concept.

Merchandising team guides the designer to identify hot selling trends in terms of styles, colors and heel heights and provide feedback. This enables designers to understand the target consumer group.

Last development

The last embodies complex shape without any straight lines and greatly determines the fit of footwear in accordance with its design, shape and volume. The anatomical information of foot must be implicitly represented by this footwear replica.

The last is considered to be the soul of a shoe. The word literally means footprint.

The last serve as the foundation for the footwear's design and construction. The shape of the last influences the fit, comfort, aesthetic appeal, functionality and construction of footwear. The pivotal role played by the last in footwear industry becomes obvious from the common phase used within the footwear industry i.e., The Last Comes First. Any deformity in designing last will undoubtedly be passed onto the footwear and perfect size and fit of designed footwear becomes unachievable. A correct, high quality last that conforms to the size, style and fit requirements in the finished footwear is thus a mandate to produce a high-quality shoe.

Footwear designers work on the concept of form follow function as the form or last will ultimately decide the fate of final footwear. Custom or better fitting shoe last is prerequisite for footwear customization. The solid 3D mold in footwear manufacturing is generally made of wood, high-density polyethylene and serves as template around which a shoe is made.

Intended design and functionality are tailored based on up to thirty-five measurements that go into the last development. As many lasts are required as are the sizes and styles of footwear being offered as an assortment for upcoming seasons. Bespoke footwear requires up to thirty standard measurements like tread, flare, toe spring, throat opening, heel-to-ball, and four separate girth measurements (ball, waist, instep, and heel). The measurements vary for different individuals challenging the shoe manufacturers to tailor their products in a manner that scales while they can still meet demands of consumers. Consequently, majority of the manufacturers design their own signature lasts that are used for production of ready-to-wear collections.

Traditionally, wood and plastic have been used for last making. The bespoke shoemakers also known as cordwainers initially used hard carved lasts composed of hardwoods like maple or beech. However, the manual method was time consuming and pain stacking. Therefore, last development has witnessed revolutionary evolution as the lasts are now available in variety of materials like high-density recyclable plastic, polyethylene, aluminum which are suitable for mass production. Moreover, metallic lasts are also becoming popular as the metal can be melted when the last is no more in use and the material reused again for creation of last of another footwear size and style. Lasts are generally created with different hinges that makes their removal from the finished shoe easier. Simple production lasts are used for shoes with a wide opening while last that can be folded out of the shoe or taken apart are generally used for shoes possessing tight or small openings.

The process of developing a last can be initiated with an existing last or footwear along with detailed account of modifications required in the last as far as width, toe shape, etc. are concerned. The new last can also be developed as the buyer furnishes a detailed list of the footwear style to be produced. Thereafter, digital last is created using a 3D last creation software. The digital last comprises of key last measurements and 2D images. The approval of digital last by the buyer leads to production of first physical last on rapid prototype machine. The last so produced is again reviewed and approved by the buyer. The rapid prototype last, after approval is made in pair of plastic production quality lasts. The production lasts are further utilized for production of first samples.

Rapid prototyping is used for realization of footwear components and fitting in 3D before proceeding ahead and design finalization. The entire design appearance can be evaluated before last preparations thus result in money and resource saving. Virtual designs from CAD systems are used and transformed into solid objects, building them up layer by layer and gradual creation of desired shape. The shape created through this mode will be 3D and a perfect virtual simulation of the physical model.

Designer last

The lasts initially used by designers for translating their sketch into first wearable prototype is referred to as designer last. Footwear designers at the onset must consider the changing toe shapes from season to season that will influence the last designing.

The designer last which will be utilized for creation of model shoe, is created as the designer's CAD generated or hand sketched footwear model. Prototyping the look, feel and fit of finished shoe can be evaluated by the designer during this stage by incorporation of style, size and fit characteristics required in the finished footwear in the designer last.

The first last is a handcrafted wooden block however, bulk production of footwear demands multiple plastic lasts in all the sizes and styles to be manufactured. Finalization of each minute detail such as stitching, colour combination and hardware, material quality and evaluation of workmanship is targeted at this stage. Thereafter, the lasts are examined for any imperfection in the design and designer may present the finished prototype to designer director, brand, etc. seeking feedback for the design. As the prototype model has been approved, it has to be mass produced.

The designer last is thus replicated into production last to be utilized for mass production of footwear style. The development of designer and production last is a prerequisite irrespective of the brand or intended footwear style.

Creation of production/development last

A separate last referred to as production or development last is produced from designer last for each style in required sizes and widths. 50 different footwear styles in different sizes and width combinations entail fifty individual lasts. The production last is developed by duplicating designer's original last on polyethylene log using lathe machines. The development of production last is followed by creation of CAD developed flat representation to be subsequently utilized for pattern-making.

Pattern-making

The accurate and desired fit of finished footwear is achievable only if patterns for different components have been accurately cut. The last plays a pivotal role in pattern-making. Pattern maker uses the last as a guide to cut the first pattern piece for shoe upper. Modellista is an experienced technician who has meticulous knowledge of foot measurements and play with proportions while making patterns of each footwear component exactly like designer's sketch.

The pattern-making by the use of last is accomplished by wrapping the last with strips of adhesive tapes to obtain key design and reference points through pencil marking along adhesive tapes. Eventually, the tapes are cut along the design lines and peeled away from the last. The peeled tapes with design details like eyeletting on upper of footwear, stitch marking, positioning of welt or any additional trim on footwear's vamp are flattened onto pattern paper. Thereafter, patterns for all components are traced and finally cut preferably by die cutting. The pattern-making is followed by die manufacturing.

The pattern is finalized and taken into production after the necessary adjustments from first prototype fitting. Grading of patterns is then accomplished based on the chosen materials and sizes to be produced. Manual, automated or combination of both the processes can be followed for cutting multiple pieces of upper. Delicate pieces require manual cutting while larger pattern pieces can be cut with a machine.

Computerized process starts with designation of number to each skin followed by capturing photograph of shape of skin and matching to the corresponding number. After matching the skins, computer creates the pattern layout, positioning and rotating the pattern at different angles.

Laser outline of each pattern piece is placed by a single click of mouse followed by arrangement of pattern pieces that the skin can accommodate. A number is designated to each skin for which computer memorizes the layout. Skin defects are to be avoided and estimation of best parts of skin for front, side or back of shoe upper is essential. Once the pattern layout is complete, the skin is folded and transferred to cutting table.

Computer recognizes each skin from its corresponding number during the cutting process. The pattern pieces are cut in accordance with the layout which computer recalls for the skin as per designated numbers.

In the absence of high-end computerized technologies, the footwear manufacturing units utilize heavy weight press and metallic dies that are exact replicas of pattern pieces. Each die is positioned on the leather, eliminating any visible defects. An evenly cut pattern piece is obtained as the pattern cutter positions the press correctly, applies pressure automatically and blades of die cut through the material. Die manufacturing is discussed in the following section.

Die manufacturing

Dies are the metallic replicas of paper patterns obtained by wrapping adhesive tapes on lasts and subsequently tracing on paper. The number of dies to be manufactured will depend on individual components and thus patterns required for a particular footwear style. The dies are capable of cutting several fabric layers simultaneously however, when leather is used for footwear construction dies cut through single or at the most two leather plies. Individual dies for each component of shoe are prepared by bending sharp steel strips around the paper patterns.

The manufactured dies are thereby utilized for cutting patterns required for a particular footwear style like men's dressy upper. The dies after positioning over the materials are pressed to cut through lay composed of several fabric plies. Imperfections are avoided in leather by manual cutting of individual layers by skilled craftsmen for efficient material optimization. Apart from die cutting, other specialized cutting tools are used in footwear industry for pattern-cutting like straight, round, band knife especially when the footwear are primarily composed of textile materials.

As the patterns for upper are cut by any of the methods discussed above, any design detail to be added as per designers' proposed design is added while the upper is in flat form.

Design detailing on the upper

The operations for applying design details on upper of footwear vary as per the footwear style. Any design detail to be included should be executed on flat upper pattern pieces before sewing and lasting as the process becomes tedious on a three-dimensional shape after lasting.

Wingtip men's footwear requiring punching design should be punched on upper as the flat pattern pieces are cut.

Marking for alignment of components

Marking proper closing of the footwear requires identification marks on upper to align with markings on the soles. Permanent or washable marking can be used during this stage. Permanent ink or metallic foil is used to stamp size, brand name, style and place of manufacturing. Metal dies are made for the stamps. Stitching lines, alignment can be identified by temporary marks applied manually with a template using chalk-based ink, silver ball point pen. Temporary marks should be visible only during stitching and should disappear before finishing and shipment of footwear. Thus, fluorescent ink visible under UV lamp is used for temporary marking during the process.

Edge treatment

A thin layer of skin is removed to reduce the thickness along the edges of upper by a process called skiving. Thereafter, clean finish is imparted by subjecting the edges to burnishing, binding and folding.

Assembling of upper components

The edge finishing of upper is followed by stitching together of all the upper components. However, a general practice is to glue the edges of patterns prior to sewing to maintain convenient and faster sewing process. Glue is applied along seam allowance for holding separate pieces together. The disadvantage of glueing process is that it binds the leather making it less breathable.

Individual pattern pieces required for a particular shoe style are sewn together in fitting room. The process includes stitching vamp lining to vamp, quarter lining to quarter, attaching quarter to vamp. The stitching at this stage is accomplished by stitch marking to locate accurate position of upper and other components. Comfort is engineered into footwear by process called doubling which involves placement of fabric interlining between the upper and lining. The interlining in contact with foot's skin should be breathable, skin friendly so as to ensure quick sweat dissipation from wearer's foot. Furthermore, eyeleting and hole punching for lace insertion is carried out for styles that are fastened by laces like oxford, wingtip, etc. Upper pieces must receive the stiffeners, toe puffs and reinforcements before stitching. The upper after being attached with all reinforcements are sent to sewing machine operators for subsequent operation. Flat bed, post bed and cylinder bed sewing machines are deployed for stitching the footwear components. Manual or computer aided machines can be used for accomplishment of this operation. A fixed sewing head on a pre-programmed auto-stitching machine can perform all the stitching operations. The upper components are preloaded and machine executes lockstitch, chain stitch or any stitch pattern as per design requirements.

The last of the operation to be performed in fitting room is functional stitching e.g. to conceal the construction details, or decorative top stitching with contrasting colored threads.

Lasting

Apart from pattern-making, the designer or production last is crucial to provide formability and required shape to the shoe upper. Accordingly, lasting operation is performed in which assembled footwear upper is pulled over the last to ensure that it conforms to the shape of the last. The process is accomplished by molding the finished upper onto last and attaching to the sole. The duration for which upper is held on the last is dependent on manufacturing process and weight of leather. The pull-on the upper should be evenly distributed to avoid wrinkles in upper or lining which may hamper the appearance or the footwear comfort. The desirable quality, style, appearance and functionality of footwear determine the last method to be employed. Irrespective of the lasting method, the general procedure followed for footwear construction is closure of heel area followed by closing the top section and eventually closing of the sides.

Two methods are generally employed for lasting – flat lasting and force lasting.

Flat lasting - upper is positioned on the last, pulled down followed by its attachment to insole at toe, side and seat of the last.

Force lasting - upper is closed with allowance and given a bag like structure using California or moccasin construction. Subsequently, the last is pushed into this bag formed by upper. Alternatively, the bag can be slipped over the last and finally finished. The method is suitable for light, soft and flexible upper materials as it requires accurate patterns that can fit the last accurately.

The lasting method to be employed for joining the upper and lower components of footwear is chosen based on quality and price point of footwear, its functionality and appearance. Lasting techniques frequently used in footwear industry are Goodyear welt, Mackay lasting, Littleway and cementing.

Goodyear welt

A firm attachment and smooth insole is ensued as the upper and sole are attached by Goodyear welt. The technique is a machine-based alternative to hand welted method. The construction involves application of two seams for attachment of shoe bottom. Welt, upper, lining and insole are held together by a concealed chain-stitched inseam. Outsole is attached to upper by lock stitched out seam. Good year Welt is a popular lasting technique among all genders specifically when sturdiness and medium to heavy weight constructions are demanded. Moreover, the construction renders footwear strength, durability, comfort and bulkiness. The construction is identifiable by leather welt surrounding the footwear and stitch going through the welt and the outsole. Female's footwear, active sports, street shoes and men's dress shoes primarily use the aforesaid construction technique.

Cementing is the least durable lasting technique as it involves attachment of sole to upper by means of adhesives rather than stitching. However, the construction method is simple, quick, cost effective and flexible method. The upper is shaped, pulled onto last and gets the sole attached without the use of welt or stitching. Cementing group includes staple lasted, tack lasted and skeleton insole cemented shoe.

Cementing is generally preferred for casual shoes, sneakers, chukkas and other rubber soled footwear which do not require stitching. The disadvantage associated with the aforesaid lasting technique is the inability to resole cemented footwear.

Mackay also referred to as Blake lasting is accomplished by tacking, stapling, cementing and stitching. The less durable methods like cementing, stapling are deployed for attachment of upper to insole. Thereby, the outsole is attached to assembled upper with a chain stitched seam. The construction is used for dress shoes however, it is not as bulkier compared to Goodyear welt construction. The construction is distinct from other lasting techniques in containing a stitching line inside the footwear following the entire edge of footwear. Stitching goes through a groove which is made into top of the outsole.

Little way lasting is the stitch down construction involving direct stitching of sole to the upper in which sole stitching is visible on the foot side of the insole and on the outsole bottom. Abrasion of the stitching is minimized by a concealed seam in a groove on the sole's bottom. Little way lasting technique is accomplished by direct stitching of sole to upper.

Slip lasting also referred to as moccasin shoe construction is primarily used for constructing light weight running or racing shoes. The mesh and polyester upper are left long and patterned so that they meet in the middle of the shoe. The sides are joined next followed by steam heating of upper. The upper is pulled onto last with footwear being laced up tightly. The upper is cooled to tighten it before the bottom being attached to the upper.

California or Strobel is a popular construction for casual and athletic footwear. The process involves stitching together of upper socks and wrapper of footwear before forcing upper onto the last. The final gluing and lasting is followed by cementing of sole to achieve a strong and sturdy finish. The construction provides cushioned feel to wearer's feet.

Stitch down lasting Stitching goes through upper on top of the outsole in Stitch down method which is considered to be traditional method of shoe construction. The lasting edge is turned outside over the insole and uses a vertical lockstitch to join the upturned portion onto insole or midsole. Thereafter, cementing is used for attachment of outsole and heels.

Assembling remaining components

The footwear is completed by assembling remaining components as the upper is assembled and lasted. Assembling process involves attachment of counters, sock linings, outsoles, shanks and heels. Stiffening materials strengthen the back part of footwear while sock lining conceals seams and stitches used for sole attachment.

Shanks are supporting, reinforcing steel pieces and sit between shoes' insole and outsole. All the components must be handy during assembling to ensure an uninterrupted, smooth process.

Bottoming

Bottoming is the subsequent process carried out in bottoming room whereby the shanks and fillers are attached to the lasted upper. Furthermore, the outsole is permanently attached to the shoe upper. The footwear is completed as the insole receives the heel by nailing.

Finishing

The bottoming of footwear is succeeded by finishing operations that render smooth, shiny appearance to shoe upper as well as the outsole. The finishing operation concocts the footwear for packing and finally shipment to retailers. The footwear finishing involves buffing, polishing, treeing, spraying, lacing and trim attachment and eventually the final inspection before sending the footwear to packing room.

Finishing is accomplished to render the upper as well as outsole of footwear velvety and smooth feel.

Soles and heels receive mechanical, chemical operations during bottom finishing while upper finishing is done by chemical finishes and shoe dressing.

Shaping the soles and heels to correct dimensions is essential for bottom finishing. Accordingly, any excess, faulty material containing scratches, blemishes are to be removed. Fine grade, abrasive is used to polish the soles. Thereafter, a shiny finish is imparted to soles by wax application. Matching or contrasting color can be imparted to edges of soles by painting with ink.

Upper finishing involves cleaning for dirt, scratch marks removal, filling of micro cracks, conditioning of leather, buffing, brushing and polishing.

The operations performed during finishing before making the footwear ready for delivery to retail outlet or customer and can be classified as buffing, polishing, treeing, spraying, lacing and trimming attachment.

Buffing involves the use of fine sandpaper on the footwear's bottom to render smooth, velvety appearance to the shoe's bottom by rubbing it with fine sand paper. The upper of footwear is imparted luster by polishing with wax.

Bottoming and polishing are followed by treeing which involves placement of assembled footwear onto the production last to ensure that the final footwear exactly conforms to the shape and style of last.

Thereafter, lustrous solution is sprayed on the upper to bond the fibres of upper material, preserving and protecting the finish. Thereafter, as per the footwear style any additional components like laces, braids on oxford, wingtip, velcro on sneakers and straps and buckles on marry jane are added. Likewise, ornamental components or trimmings like rickrack, pompom, tassels, and embellished stones can be attached on footwear like thongs, espadrilles.

The last operation which marks the footwear ready to be packed and shipped is the final inspection to ascertain that the finished footwear meets the standards and norms set by the buyers and retailers as discussed in the following section. The inspection of footwear is followed by its packing in boxes and can be shipped either to warehouse for storage or directly sent to footwear brands, sellers and retail outlets.

Evaluation & label attachment

Evaluation of the final finished footwear is essential before shipment to retail outlets. Inspection of the finished product is mandatory at eventual stage to ensure that the developed footwear is constructed in accordance with brand/ consumers' specific sheet.

The test procedures to be followed will vary as per the footwear styles. X-ray machines are used by some footwear manufacturing units to check for correct placement of pins, staples on high heel footwear and ensure that heel will neither collapse nor lose its shape.

Tear strength, extension, break, stitching strength, breaking stress and fastness to rubbing are other important properties which need to be evaluated for footwear.

After evaluation of footwear, they receive labels, tags and are finally wrapped and packed as per requirements for shipment laid by brands and retailers.

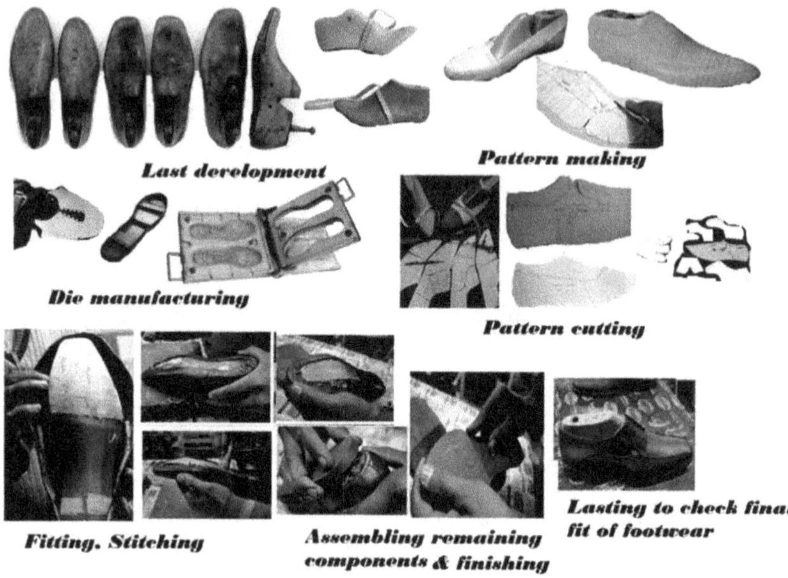

Figure 4.5 Footwear construction

4.1.6 CAD intervention for Virtual designs and fit customization

The traditional method of footwear construction has been discussed in the previous section in details. The primitive methods of footwear construction involved development of a pair of trial shoes. Although the process was time consuming, resulted in wastage of expensive, top-quality leather and cost ineffective however, construction of trial footwear was undertaken to curb slightest possibility of an ill-fitting pair of footwear for consumers. The pairs of trial footwear were then assessed for their fit and any misfit was rectified by adjustments made to the last by marking the points that required building up or reducing. The trial shoes were however, useless after the process. The quest of consumers for footwear's style and fit personalization can be addressed by a more innovative and time efficient CAD technology which can offer high-quality customized product at affordable prices and faster pace. Accordingly, the demands of fashion conscious and inquisitive consumers can be gratified by involving consumers in the design and selection process in the virtual design, customization and trial systems on the software thus reducing the production costs and time towards development of trial footwear. The footwear designs created using CAD technology are based on the 3D foot scanning data;

thereby enabling both fit and design customization. Consumers are offered the flexibility to choose both fit and style customization before the development of shoe-last as per size, style and fit specifications. The foot scanner enables the scanning of consumer's feet which can be saved as digitized foot. Furthermore, consumer can choose the style of the footwear from array of footwear styles already available in the database. The selection procedure is followed by creation of desired footwear of the particular consumer as per his/her feet's shape and size in accordance with the scanned feet data. CNC machine is utilized for the development of footwear last. A unified system provides access to the footwear design for consumers as well as manufacturers. Specifications of shoe-last styles are created by manufacturers for presentation to consumers. The feedback of consumer can serve as valuable inputs to modify the design as per customer's needs. As the customer customizes style in the system, the normal sizes or fit can also be customized. The customers and manufacturers are linked in the system and both play a crucial role in the production of a desired shoe-last. The footwear lasts are designed according to heel height, toe shape and footwear type. Furthermore, fit customization is essential to generate a size appropriate for a specific consumer and thus calls for sizing and grading for shoe-last customization and fit evaluation.

4.1.7 Men footwear styles

Brogue is a popular male footwear style characterized by decorative perforations on vamp and wingtip lace ups.

Loafers are slip on type of footwear featuring low heel and absence of any fastening components like zippers, lacings or velcro. The footwear styles are equally popular among males and females owing to ease in wearing and taking off footwear and has become synonymous with casual look and comfort to the wearer (Figure 4.6)

Mocassins originated as footwear for American natives but is no longer restricted to them and is globally a popular footwear style. The footwear are generally developed in soft leather with sole and side made of one leather piece and top comprised of an additional leather piece of same or different color as the sole and sides. The look of mocassins is completed with decorative embellishments like leather fringes or beading appearing on the vamp.

Monk is traditionally considered to be men's footwear however, the style is trending among females and kids as well. The footwear style features a buckle on the upper without the inclusion of any other fastening element like lacing, etc.

Oxfords are considered to be the finest, lace-up dress men's shoes that were once popular among students at Oxford University however, the footwear style available generally in black and brown leather has gained popularity among men of all age groups as formal wear. The footwear style features closed lacing and two narrow rows of stitching along almond shaped toecap. The closed lacing system with lacing eyelets concealed underneath vamp renders snug fit and clean finish to oxfords. Variation of classic oxford is the one-piece oxford that rules the fashion arena. The latter is constructed of single leather piece instead of various pieces sewn together. The leather pieces are sewn as a single piece providing sleek and sophisticated look to the footwear.

Several affordable, premium and luxury brands namely Aldo, Clarks, Topman, Kenneth Cole, J.Crew and Crockett & Jones, George Cleverly cater to the ever increasing demands of oxford lovers globally.

Derby/Gibson also known as open lacing footwear is a men's casual footwear style with lacing eyelets stitched on the top of vamp. The laced flaps cover a one-piece vamp/tongue as additional panels sewn onto the shoe in contrast to oxfords which are devoid of any flaps and eyelets for laces are present on the shoe itself.

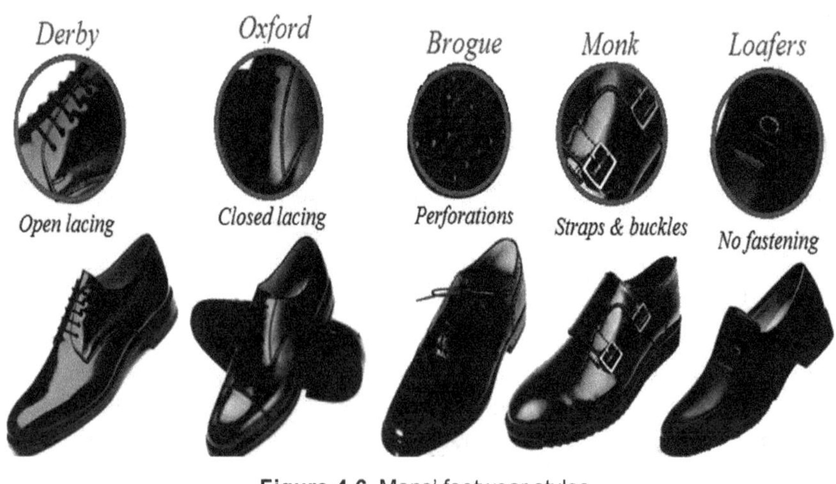

Figure 4.6 Mens' footwear styles

Trainer/Sneakers also known as athletic shoes consist of a rubber sole with lacing over the instep. The footwear style is referred to as trainers as they are preferred by athletes and sportsmen during their sports training. An athletic shoe can also be described as a tennis shoe. The sneakers are primarily designed for sportsmen however, the quest for comfort by consumers in normal wear prompt footwear designers to design sneakers with an essence of minimal decorative embellishments for low activity level liking walking, jogging or for travelling when the consumers wish to seek casual, sporty and comfortable look (Figure 4.7).

Figure 4.7 (a) Trainers & Sneakers for leisure & active sportswear

Figure 4.7 (b) Trainers & Sneakers for leisure & active sportswear

Espadrille is a unisex footwear style featuring sole of raffia or jute and an upper composed of leather, suede, canvas fabric and is an ideal casual wear. The earlier styles contained a canvas strap wrapping around the ankle however, in recent times, espadrilles have witnessed significant variation in fastening elements as shown in Figure 4.8. The female espadrilles features high heels, platforms, wedges or flat, fastening extending up to ankle with decorative embellishments and trims beautifying the footwear unlike the men's espadrille which are comparatively simpler in design and construction with minimal trimmings.

Figure 4.8 Male & female Espadrilles

4.1.8 Women footwear styles

A myriad of female footwear styles dominates the fashion world. The footwear designers fully exploit the female buyers' shopaholic attributes and meticulous selection of the accessories to complete their look. The female footwear styles differ mainly based on materials and trimmings used for their construction, toe cap shape, vamp and quarter construction that determines the exposed or covered foot area, fastening elements and heel types.

Boots are one of the footwear styles that can be functional as well as a means of adornment and embellishment. Knee or thigh length leather boots creates a style statement and complement the look of western attire like a shift or sheath dress. Boots provide insulation and warmth to feet in cold environment, protection to feet against external elements such as water, snow, mud, etc. Furthermore, the pivotal role played by boots in varying sporting activities like riding, skiing, snowboarding, ice-skating cannot be undermined as they provide complete coverage and protection to feet in wet, damp or extreme climatic conditions. The level of insulation, protection and comfort dictates the coverage provided by boots. Accordingly, boots can provide complete coverage to feet, can extend up to ankle, knee, thigh or even hip. The boots can be distinguishable based on the materials used for their construction, fastening elements, their length, height and shape of heels. A variety of boot styles like Chelsea, dress, cowboy, go-go, hessian, mukluks, riggers, Ugg and valenki boots rule the footwear industry.

Chelsea Boots, dealer boots or paddock boots are tight-fitting, ankle-high boots that were primarily used for horse riding. The boots feature elastic siding extending from just above the welt to the top of the footwear. Chelsea boots first made their emergence in Victorian period and have more recently made a comeback in fashion arena.

Dress boots are short boots extending up to ankle, generally composed of leather or imitated leather like suede. The boots primarily find application as formal wear or in adverse climatic conditions and outdoor situation to provide protection to feet.

Cowboy boots as the name implies were initially popular among cowboys and are classified as riding boots. The salient features of this boot style include cuban heel, rounded to pointed toe, high shaft with no lacing for fastening. Exotic skins like alligator, snake, ostrich, lizard, eel, etc. as well as cowhide leather are generally preferred for construction of cowboy boot's upper. The boots can further be classified as western/ classic and roper boots.

Fashion boots unlike hiking, riding and rain boots don't have any functionality associated with them; however, the women footwear style is popular among fashion connoisseurs for making a striking fashion statement and complements the look. Fashion boots can extend up to ankle or can even be thigh length and compliment well with formal, casual and corporate looks alike.

Hessian boots were primarily worn by military personals; however, the footwear has been gaining popularity among civilians as well. The boots are characterized by low heel, and a specially designed semi-pointed toe. The boots extend up to knees usually embellished with a tassel at each shaft's top portion.

Rigger boots are considered to be work boots or pull-on safety boots that were primarily designed for workers deployed in North Sea's offshore oil rigs.

Go-go boots are a low-heeled style of women's fashion boot designed in variety of materials like leather, suede, etc.

Mukluks or Kamik are soft boots designed with reindeer skin, sealskin and fur. The footwear style had origin from Arctic natives who wore mukluks to protect their feet and keep them warm in extreme cold weather conditions. Mukluks follow recurring cycle with the styles reintroduced in latest times using imitated leather versions, fur, trimmings and such components that provide protection and insulation to wearer's feet in extreme cold weather.

Russian boots are calf- or knee-length women footwear that first made appearance in fashion arena in the 20th century. The boots were popular in 1930's and gradually declined in popularity however, made a huge comeback around 60s as low-heeled boots inspired from footwear worn by folk dancers.

Figure 4.9 (a) Women footwear styles

Ugg boots also referred to as sheepskin boots are developed from twin-faced sheepskin having fleece on the inside, a tanned outer surface and a synthetic sole. The materials and construction of footwear make them ideal choice in winters as the fleece on the inside provide thermal insulation to wearer's feet. Ugg boots originated initially as utilitarian footwear for providing warmth to wearers in cold climate. Gradually, the boots became popular among surfers however, the aesthetic dimension was introduced as the style recurred in 90's and 2000.

Valenki valenok are felt or wool felt boots traditionally used in Russia as winter footwear. The boot styles are not water-resistant and thus require galoshes to be worn over them to prevent water penetration and protection of outsole from wear and tear. Galoshes are rubberized boots also known as gumshoes or overshoes generally slipped over valenki as the outer protective covering against water, mud, etc.

Footwear can also be distinguished on the basis of heel types they feature. Accordingly, the footwear can be classified into pumps, stilettoes, wedges, block, platform, tapered, blade, kitten heels and ballerinas based on types and shapes of heels they feature.

High-heeled footwear like pumps (court shoes), stilettoes can have heels ranging in height from 3 inches to up to 5 inches or even more. The footwear styles are generally preferred by short heighted females to give illusion of tallness and slender legs. The heel of the wearer's foot is significantly elevated than the toe which can be a woe and trouble for the wearer due to uneven distribution of pressure on foot with maximum stress at the heel portion. The risks associated with high-heeled footwear are innumerous and the footwear is not suitable for comfort lovers. Wearing high-heeled pumps, stilettoes not just result in foot and tendon pain but can aggravate the situation further by increasing the likelihood of sprains and fractures, can create foot deformities and also agitate lower back pain.

Therefore, footwear styles like wedges, platforms, ballerinas are a better option when the wearer intends to wear the footwear for long durations and pay heed towards functionality and comfort rather than mere aesthetic appeal. Platform footwear are designed with heel and toes raised to equal amounts resulting in uniform pressure distribution on wearer's feet unlike platforms, stilettoes.

Mary Jane is a famous footwear style that closely resembles footwear worn by school going girls featuring a strap and buckle running horizontally midway along the foot. The strap across the instep along with fastening element like button or buckle at the side ensures firm grip to the wearer's feet. Twin bar and triple bar mary janes are summer wear footwear that uses breathable fabric instead of leather, suede with the toe cap opened slightly making them ideal for casual summer look.

T Bar footwear gets its name from the T strap rising vertically up in the front portion of foot. The strap has a loop or slot at the top that enables holding the instep strap across the foot at right angles. T strap stilettoes, pumps, platform and sandals are popular among females to flaunt their chic or preppy look.

Sling back is another illustrious female footwear style that has several formal and casual alternatives like a sling back thong, peep toe sling back, sling back espadrilles, etc. A sling back irrespective of heel shape and height and materials, possesses a strap around the upper part of the heel at the back, with rest of heel exposed.

Ballerinas are footwear styles that are reminiscent of a ballet shoe. The footwear style is generally flat and has a heel height of 3-8 mm.

Pumps are high-heeled slip-on footwear in which topline holds the footwear on the foot with coverage to toe by vamp, sides and back by quarters and counters providing firm grip to feet. Thus, the footwear does not contain any supplementary fastening elements, straps or lacing.

Knee high boot as the name implies are knee high boots and compliment well with miniskirts. The boots can be designed with leather, suede, vinyl with platform, block or wedge heel.

Figure 4.9 (b) Women footwear styles

Mule is unisex backless footwear style characterized by absence of any fastening or gripping element. The footwear can be peep toe, opened or closed at the toe with the back completely open. The vamps of mules are constructed of one or more pieces and may contain a strap, tassel or minimal trimming for aesthetic appeal.

Sandals are generally casual footwear styles that are preferred by wearers in warmer weather. The footwear ensures comfort to the wearer as it is an open shoe exposing most part of the foot including the toes. The straps at the center or back are fastened with Velcro for firm grip to feet.

Thongs are open sandals that are held in foot by means of thong between the thumb and second toe of foot.

Ankle boots are popular female footwear that provides complete coverage to feet and usually ends above the anklebone and below the lower calf.

Ghillie footwear closely resembles a ballet shoe and is generally used by Irish and Scottish dancers. The footwear is designed with leather, suede, vinyl and features a crisscross lace, ribbon or elastic running across the top of the foot and around the ankle.

Figure 4.9 (c) Women footwear styles

4.2 Hats

4.2.1 Introduction

Millinery is category of fashion accessories that can perfectly frame the face, enhance and complement wearer's look and can serve as protective shield against external elements. In the early days, hats were worn only for functional reasons.

Hat designing is a diverse, thought provoking and a creative discipline.

Designers need to survey and scan the mood of the season, brand signature, requirements of target consumers, market level and price point while designing for a brand/ label. Furthermore, technical considerations relevant to manufacturing and production methods also demand consideration as the hat designing is initiated.

4.2.2 Hat components

The primary components of hat include crown, brim, head band and trim. However, the presence or absence of either of these components, their shapes and materials depend largely on the hat style. Some hat styles like beanies are brimless headwear, while others may have an elaborate, wide crown as in cowboy hats.

Likewise, headband for firm and secure gripping is tied around chin in bonnet while turban does not require a headband.

Crown of hat is the top portion of the hat that covers major head area.

Brim is an upturned, downturned or straight projecting edge of hat.

Headband is a strip of leather, suede or any durable and comfortable fabric on the inside of hat that encircles the lower crown.

Decorative components are trims that ornate and adorn the hat's surface and are more often attached on the crown portion of hat. A range of decorative components namely ribbons, feathers, flowers, braids, stones, laces are widely used in accordance with hat style.

The millinery construction process initiates with designing whereby designers creates a detailed sketch incorporating all the details pertaining to style, design, trimmings and stitching. A preproduction sample which can be modified several times before reaching the final sample can also be developed using designer's sketch.

The aesthetic appeals, functionality, wear ability, proportion and price point are some of the critical factors to be considered by designers while designing hats.

Designers can simply create line of casual hats or experiment with millinery couture or model millinery that is hand sculpted intended for elite, customized look to wearers.

The next section gives a detailed description of construction method of hats in accordance with the used materials.

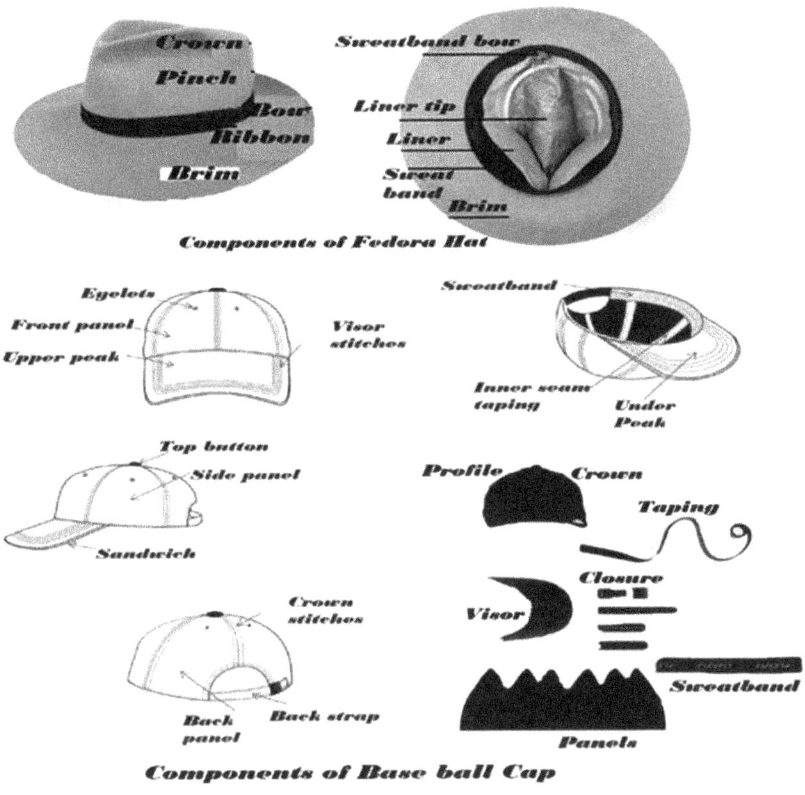

Figure 4.10 Hat components

4.2.3 Materials & Construction methods for hats

The price of the finished product, material used for the hat body, decorative trimmings and appearance dictates the construction methods employed for hats. Finer, upmarket hats are generally hand-crafted while automated methods are employed for inexpensive hats. The hat construction irrespective of material types begins with hand or CAD/CAM sketching of hat style followed by translation of sketch into prototype sample. Design of hat should include all detailing pertaining to body and ornamentation. The evaluation of the model ensures that prototype can be taken for bulk production.

Weaving, knitting, felting, crocheting and braiding are some of the commonly used techniques for millinery production. Stiffening net i.e., buckram, hat linings, trimmings and embellishments can be utilized irrespective of the technique employed for hat construction.

The construction method employed for hats is greatly influenced by raw materials, their properties like shape retention, durability, stiffness or softness and the price point of the finished hats.

Fur fibres, felts, fashion fabrics, feather and straws are commonly used materials for hats. Lightweight fabric and straws are preferred for spring/summer hats while leather, suede, furs, felts, pile and heavy weight fabrics are the fabrics of choice for fall/winter hats.

Jute, palm fibre, panama, seagrass, wheat, sisal and toyo paper are the generally used straw types for hats.

Felt, straw, leather, soft fabrics like velvets or velour, knitted and crocheted fabrics are the preferred materials for hat construction.

Selection of lining fabrics for hats is dictated by hat style and their utility. Light weight cotton linings are preferred for summer hats while wool, acrylic and felt lining may be materials of choice for winter hats and headwear.

Felt hat construction

Felt hats popular among both genders alike are generally constructed by shaping the fabrics over cones. The crown, body of the hat and brim are formed as a result of repeated steaming under pressure. Thereafter, the brim is trimmed to the desired width after additional steaming and ironing. The cone used for shaping felt hats is created by carding the fibres into a loose mass and winding around wooden cone. Crown and brim are formed as the cone is stretched over a frame.

Felt hats are finished by hand sewing millinery wire in an overcast stitch around the edge. Raw brim edge is finished by machine or hand sewn Peter sham ribbon. Wooden block or aluminum pan forms are used for providing desired shape to hats. Apart from manual blocking that is used for mass production of hats, hydraulic blocking is carried out after the hand blocking or machine method. Hydraulic blocking is accomplished after traditional hand or machine blocking techniques. The process involves application of intense pressure so that felt or straw hat conforms to the pan's shape.

Straw hat construction

Straw hats are quite popular as casual and trendy headwear. Straw is woven into three forms to construct hats. Cape line, cone and braid weaving are preferred methods employed while using straw as the main raw material for hats. Wide brimmed hats with one-piece crown and brim are woven using cape line weaving.

Specially crafted wooden forms known as blocks are used for shaping hats made of woven mat straws. The block is exact replica of desired hat style. The straw is subjected to steaming while still on the block to attain the shape of form. The hat retains its shape by addition of stiffening materials like buckram or interlining post blocking and drying. Plaiting is another construction method used for straw hats. The method involves overlapping narrow strips of straw and attaching by sewing to attain hat shape as per the required style. Thereafter, the plated material is subjected to blocking and stiffening to retain the hat shape.

Unstructured hat construction

Unstructured hat models use soft fabrics like velvet, velour, satin and the construction method is similar to that utilized for garment construction. The material is cut according to required size and shape thereby draping and sewing into soft, unstructured style.

Knitted hat construction

Hats particularly intended for winter wear like beanies and skull caps utilize knitted, crocheted fabrics for their development. Hand knitting, circular or flatbed knitting can be used for knitted hats featuring range of colors, stitch patterns and textures.

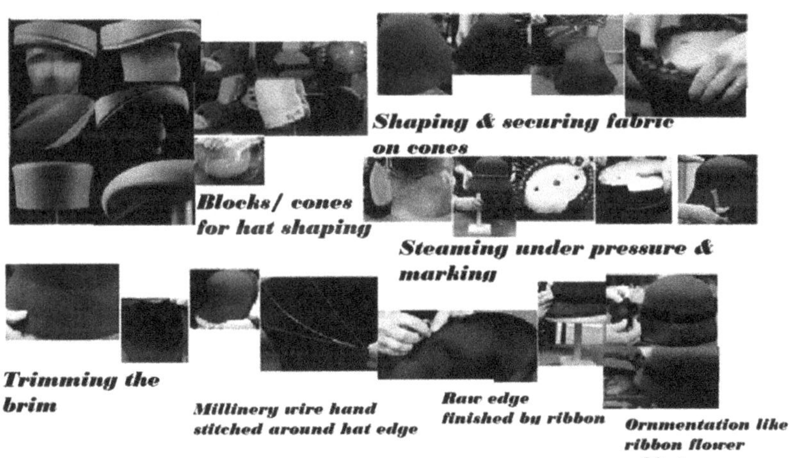

Figure 4.11 Steps involved in felt hat construction

Brimless hat construction

Conical shaped, small-brimmed or brimless hats are created using cone weaving while braiding technique is used for overlapping narrow strips of braids from crown to brim portion. As the hood that will form the hat's crown is woven, the crown shape is used to secure the damp and stretched material over it. Retention of hat's shape is possible by thorough drying of straw and coating with thin layer of varnish. The crown is the first part to be blocked followed by brim. Bias tape or Peter sham binding is used for edge neatening and finishing the brim edges. The tapes are narrow woven fabrics possessing certain level of stretch and are usually folded or stitched around the raw edge.

Figure 4.12 Straw hat types & construction

4.2.4 Hat styles

A myriad of hat styles are designed as per consumers' preferences, functionality or level of protection, dressing ethics, professional and end-use requirement.

Baseball caps available in suede, canvas, felt and leather are indicative of a particular profession, sports or college team. The caps are one-size fit all featuring an adjustable band in the back and rounded, fitted body composed of triangular inserts or gores finished with a peak at the front. A variant of baseball cap is a bejeweled cap embellished with trimmings like stones, pearls, beads to render funky, dapper look.

Beanie is a head snugging, brimless hat characterized by triangular gores and is preferred during winters or for casual usage.

Beret is a unisex hat style that is soft, unstructured model and does not require blocking during its construction. It is a small brimless cap with flat crown and generally worn on an angle at the top of head.

Cowboy hats also referred to as ten-gallon hat features a high crown and broad brim. The preferred choice of material for cowboy hats is felt due to easy cleaning ability, durability, moulding ability and availability in an array of thickness and colors.

Derby also known as bowler is a stiff felt hat featuring rounded crown and narrow rolled brim.

Fedora or homburg is soft felt hat with creased crown and brim rolled upward on the sides.

Panama is a plated straw hat featuring high crown with a centre crease.

Pillbox is a woman's hat style that features small, round crown and is devoid of brim.

Bonnet is a brimmed hat style popular during sports like golfing, horse riding and features a ribbon for tying under chin for firm grip for wearer during dynamic sports. Fastening under chin prevents the hat from falling down as the person is involved in active sports. Moreover, the sturdy structured hat provides protection to wearer's head in event of fall, collision against heavy, sharp objects.

Alpine is a man's soft felt hat style featuring turned up brim on one side.

Turban is generally composed of draped fabric fitting around the head and is soft constructed headwear. Turbans are religious symbolism for Sikh cult. However, turbans of soft, silky drape able fabrics are very much in fashion among females adorning ethnic look.

Top hat is a formal men's wear usually in silk and composed of tall crown and rolled brim.

Breton is a woman's hat style characterized by small evenly rolled brim.

Cartwheel is a female large sized hat featuring wide, circular brim.

Sailor is a low crowned hat with a stiff brim.

Garden party is a straw hat style with a wide brim embellished with decorative trimmings namely ribbon, rick-a-rack, appliques, flowers, etc.

Ski cap is a knitted headwear characterized by wide, turned back cuff and usually pompom on top.

Slouch is a water repellant, rain hat usually made of canvas or other durable material treated to render water resistant properties. The hat consists of center crease and a turned down brim.

Hood is an extended head covering attached to back of t-shirt, sweat shirt or knitted outerwear preferably of same material or contrasting colors.

Profile is a woman's brimmed hat style featuring a brim that is turned on one side and downward on the other.

Fez is a unisex hat style distinct from other hat styles due to high, cylindrical shaped crown completed with a tassel at the top.

4.3 Handbags

4.3.1 Introduction

Handbags once considered to be functional entity have evolved as an indispensable accessory in the fashion business. Selection of handbag is dictated by status and co-ordination with attire and footwear; however, the fit and suitability with an individual's life style are equally crucial. Handbags are relegated as sturdy and functional equipment that support busy and diverse lifestyle of handbag users.

Brands like Hidesign, Caprese, Baggett, Gucci, Guess, and Aldo are pioneers in handbag designing that match the demands of modern era's effervescent and flamboyant consumers particularly females. Furthermore, brands focus on coordinating the entire look by designing the product mix so that design theme, mood of the season, color and print are reflected in all the merchandise i.e., apparels, footwear and handbags.

Accordingly, each selling season, fashion brands like Zara, Marks & Spencer, H& M, Vero Moda present an assortment of apparels and accessories that are coordinated enabling a one stop shop convenience for fashion buffs. Consumers while buying a handbag seek flexibility, aesthetic appeal, strength, efficiency and durability to enable them hassle free and convenient travelling or shopping experience while carrying all the necessary belongings in the bag.

Accordingly, handbag designers consider the needs of the consumers and the end usage of handbags to design a myriad of bags that offer distinguishing features to fulfill consumers' requirements.

Clothing line in each selling season dictates the need for specific handbag style. Several designer labels are proactively involved in developing classy, exorbitant handbags to cater to the upmarket segment. Accordingly, the labels have been entering licensing agreements to ensure that consumers can recognize the labels and consider the same while making purchases.

The merchandising trends of apparels are followed in handbag segment as well with the assortment of handbag styles put to display in separate stores or special departments within stores. Zara reserves store space for accessories like footwear, handbags and use impressive props and visual merchandising tricks to attract more foot traffic.

4.3.2 Handbag designing

Handbag designing necessitate the consideration of key design elements – line, balance, silhouette and proportion. It is crucial to determine bag's proportion in relation to its body. Another vital element is exploration of end usage and mode of carrying the handbag. Balance is related to harmony or discord of design detailing as well as hang of the bag on being carried or worn. The shape of filled handbag often tends to alter under the pull of gravity. Accordingly, all these design elements need to be explored and any imperfections which are expected in finished product should be resolved at 3D design development stage. Furthermore, a lot of efforts go into researching and sourcing for the appropriate type, finish, size of fitting and hardware. Handbag designers and manufacturers also need to be mindful of the weight, balance and harmony in the placement of fittings on the handbag. An oversized hardware, rivet, magnetic closure on a delicate fabric handbag not just is mismatch, visually unpleasing but may also cause sagging and tearing of the handbag's outer covering due to its bulk.

The handbag designers give precedence to season, trends in apparel merchandise and consumers requirements while sketching, conceptualization and material procurement. Color palette, materials and motifs are influenced by season and trends in apparels.

Designers needs to work meticulously on the initial sketch as a precise technical sketch is mandatory for tech packs used for handbag prototype and bulk production. Black and white sketch created by designer should include all measurements and minutest of design and construction details.

Designers come up with a variety of designs in two dimensions and once satisfied, he switches over to 3D sampling and testing of handbags.

The precise handbag construction demands creation of series of samples to evaluate material choice in relation to construction, seam types, testing details and edge finishes.

Any flaws in designing can be edited during sampling stage by the revision or adjustment in designs. Thus, it is ensured that the prototype and bulk-produced handbags are free of any typological or design errors.

Figure 4.13 Hat styles

Designer is further held responsible for development of any hardware required for handbag namely buckles closures, zipper pulls, feet, logo plates, rings, chains, etc. Accordingly, the designer does a lot of brainstorming for precise selection of metals, metal finishes and produces a black and white sketch including all the measurements and glimpse of all possible views.

Hardware manufacturers seek the designers' approval of final look of hardware.

A trim development team may also work in co-ordination with designers to work out on best possible handbag hard wares and trims for the designers' concept.

Figure 4.14 Virtual prototypes of handbags

4.3.3 Handbag components & materials

Handbags can vary in the number of components from as few as three or four to as many as thirty components. The number of components that complete the handbag are style specific. A clutch or handheld bag may not require handle or grasping element while handle is a crucial component for tote or a shoulder bag.

The crucial components of handbags are classified as outer covering, frame, side panels, lining, handles and closures.

The frame of the handbag is the body over which handbag is constructed.

Outer covering is the main body of handbag which may be supported on frame.

Gussets or expansion gores are the side panels that can increase the volume and capacity of handbags as they allow for expansion. *Gussets* are usually positioned on the bag's side.

Lining provides a neater finish on the interiors of handbags by concealing the construction stitches, glue, etc. Leather is used as lining material to impart durability to handbags however, the leather linings are expensive. Fabric interlinings and linings are available at lower price points and are more frequently used. Lining protects the handbags from stretching, accommodates additional compartments and pockets in the bag interiors. Leather, silk or colorful printed fabrics are preferred choices for lining materials for handbags. Lining, irrespective of its material should be strong, durable and composed of tight weave structure to avoid thread unravelling on handbags. Extra structure to the bag can be added by reinforced linings with a light iron-on backer. Leather or suede lining is used in luxury branded handbags while faux suede, light cotton drill, polyesters, rayon and cotton mixes are preferred choices for casual, affordable handbags.

Handles are the grasping elements that enable easy and comfortable handling of handbags. They vary in length and materials ranging from wood, lucite, beaded handles to metallic, chain link handles to those constructed from fabric, leather, vinyl, etc. Handles can be attached to top or sides of the bags and should possess sufficient strength to withstand the weight of contents carried in handbags.

Closures are other important components of handbags ensuring that personal belongings of the users are secure without risk of falling out or being stolen. A variety of closures namely buckles, zippers, locks, velcro, clasps, frog closures, snaps, rivets and drawstrings are widely used depending on handbag styles and level of security required for goods carried in handbags.

Reinforcements are important components for providing structuring and support to the outer covering of the bags. The reinforcements are added to base of bags, top edge, gusset and behind heavy hardware riveting onto bag's body. Reinforcements on gusset provide additional rigidity while that on hardware supports the bag from collapsing under bulky hardware.

Type of leather, fittings and hardware, handle styles and construction method determine the selection of reinforcement types. The areas requiring strengthening receive reinforcement (via glueing) prior to bag construction. Papers, cards, boards, fiber boards, woven and non-woven fabrics, reprocessed leather, foam, woven and sheet plastic and linings are some of the commonly used reinforcement materials for handbags.

Hardware for handbags

The types, shapes and location of closures can greatly influence the overall look of the bag. Branded metal closure or a zipper with signature zipper pull has been used by handbag designers to endorse their label or brand.

Hardware of handbags comprises all the metal components of bag like metal closures, rings, D-rings, buckles, lock and key, feet studs, studs, snaps, zipper pullers, leg plate, nylon side release strap fasteners, two-part plate locks, turn locks, dog clips, eyelets, decorative studs, press studs, cord locks, square o ring, buckled and pre-fabricated handles are the various hard wares extensively utilized for handbags. The selection of hardware is based on styles, size, materials and end usage of handbags.

Designing or selection of bag's hardware should be based on the consumers' need of security of their belongings carried in bags.

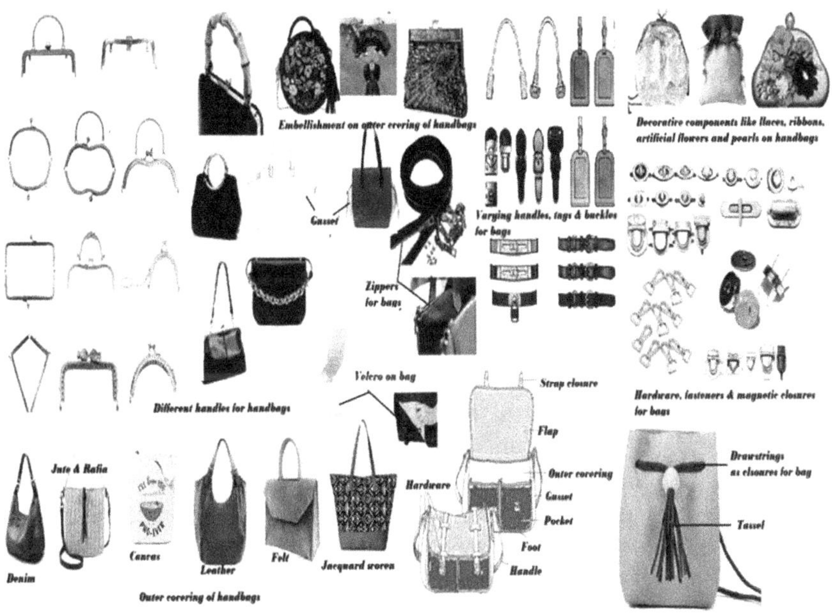

Figure 4.15 Handbag components

Feet studs are not mandatory components for all handbag styles. Birkin bag uses feet studs as a signature design element while hobo or coin purses are designed without feet studs. Wheels have become crucial components for mobility on suitcases. More recently, high tech luggage has been designed comprising of a four-wheel system equipped with swivel wheels and foot brake for easy maneuvering.

Handles - Bags can have handles or shoulder straps for carrying them conveniently. The handles come in a variety of materials ranging from same material as the bag's outer coverings or of some durable material like webbing that possesses the ability to hold weight. Handles particularly shoulder straps for additional comfort are usually reinforced or padded.

Flaps may be present in some handbag styles like messenger bag which is designed with flap. The flap is either attached or provided as an extension from back wall and falling over the front.

Bag pockets - Bags these days are provided with numerous inner and outer pockets for user's convenience to carry valuables securely. Both decorative and functional pockets appear on the handbags.

Handbag materials

The sourcing and procurement of materials for handbags is an exhaustive and pain stacking process owing to the wide range of raw materials utilized for handbag construction. The materials for handbags cannot be easily purchased at retail, but involves extensive research of sourcing suppliers, tanneries and leather merchants.

The style, mode of handling and closing the bag and its application area are crucial considerations while procurement of material for the outer covering, handles and closing elements of the handbags.

A variety of textile and non-textile materials are used for handbag construction. Textile materials like denim and jacquard woven fabrics, canvas, wool flannel, corduroy, silk shantung, sweater knits and non-textiles materials like leather, suede, vinyl and faux fur are extensively used for outer body of handbags. The textile fabrics utilized for handbag construction are generally imparted to several printing, finishing and embellishment techniques to impart functionality and aesthetic appeal to the finished handbags. Accordingly, the fabrics intended for handbag construction can be finished with functional finishes to impart water and flame repellency, can be block printed, screen or digitally printed, tie-dyed, or subjected to surface ornamentation techniques like embroidery, crocheting and applique work for value addition.

The non-textile materials utilized for handbag construction are often pleated, shirred and quilted. Natural straw like willow, wicker or synthetic straw like pontova, toyo are increasingly being used as sustainable alternative for handbag design and development.

Foam is another material used for bag interiors to covers hard edges of the frame and provides protection to the outer covering from metal frame.

A detailed account of various materials used for outer covering, handles and closing elements of handbags in accordance with their application areas is covered in the following section.

The outer body of handbag which is reinforced and supported by frame is composed of an array of textile and non-textile materials such as canvas, jacquard woven fabrics, brocades and other extra figu felt, non-woven red fabrics, satin, straw, bamboo, hemp, velvet, leather, felt, vinyl and plastic as discussed above.

Canvas, jacquard woven and denim fabrics are preferred for sturdy, durable handbags such as laptop bags, shoulder bags, totes, etc. A variety of non-textile materials like leather, suede, felt, non-woven linings and interlining are also utilized for outer covering and grasping elements like straps and handles for handbags. Wood, plastic, straw, metals and simulated stones are some other materials used for handles of handbags like bracelet, chain and shoulder strap.

The full grain, aniline finished calfskin leathers are considered to be best choices for exotic, upmarket handbag and small leather good designing.

Among synthetics, PVC (polyvinylchloride) and PU (polyurethane), the imitated versions of leather have been gaining widespread acceptance. The look of exotic leathers can be imitated by embossing or imprinting grains onto leather surface. The materials are sold by yard resulting in little or no wastage. However, the leather substitute lacks the superior handle of leather and is notorious for chemical smell. Coating on PVC and PU materials offer enhanced durability and water proofing thus making them ideal choices for sporty accessories. Nylon rips ton, polyester, canvas, and acrylic fabrics are generally coated for use in handbag construction owing to availability in an array of colors and dye ability.

Defense applications like military backpacks and hybrid bags require ballistic properties and thus uses ultra-light to heavy weight corduroy ballistic nylon fabrics. The ballistic fabrics exhibit exceptionally high durability and abrasion resistance owing to the high tenacity yarns woven into 2X2 construction. Another specialty fabric, Kevlar is increasingly used for designing accessories for military, defense personals due to fiber's inherent protective bulletproof quality.

Handbags intended for occasional purposes like weddings, parties, etc. are generally designed with embroidered, lustrous and sheer fabrics to compliment well with the attire. Accordingly, handbags for such events are designed with rich looking fabrics like brocades, silk, velvet, sateen, satin, etc. Additionally, lucite, metals and sustainable materials like bamboo, straw and hemp, etc. are increasingly being used for designing casual as well as formal handbag styles.

The aesthetic appeal of handbags can be enhanced by myriad of surface ornamentation techniques such as top stitching, label and motif attachment, appliques, synthetic stones, swarovski, shiffie, beads, fringes, tassels, laces and ribbons on handbag's outer covering.

Several printing techniques namely screen, transfer and digital printing are also utilized for customization or mass production of handbags. Many a times, closing elements like zippers, buttons, magnetic closures, drawstrings, etc. are attached for value addition and ornate look as a trimming or as functional components.

Beach bags and informal bags are usually constructed from straw cloth and straws that are available in range of fibres, weaves, patterns and colors. The materials are also available as woven into basic shopper shapes. The bags designed from straw cloth and woven straws are appealing and present a trendy and funky look however, the bags lack durability and probably last only one season. Duck canvas, Oxford and drill cottons are other textile materials extensively utilized for informal and sporty handbag development. The materials are strong and durable and available in variety of weights and finishes namely lightly waxed, heavily waxed and antiqued.

Canvas totes and shoppers are lately gaining popularity in the wake of sustainable practices. Accordingly, many handbag designers make use of waxed canvas accentuated with vegetable tanned leather straps and trims rendering traditional old school feel.

Denim, corduroy, cavalry twill and wool tweed in conjunction with leather are on designers' hit list to cater to handbag requirements of consumers from all walks of life. A variety of weaves namely dog's teeth, windowpane, Prince of Wales render unique patterns to tweed fabrics.

Likewise, an all-check pattern, herringbone and twill woven fabrics are utilized for formal as well as casual handbag design and development.

Wool felt, a non-woven option is among other materials used by handbag designers as it works well on raw-edge designs without causing fraying, curling, etc.

Felt, another popular material for handbag construction is produced from wool fibres that are matted together. Felt has a hairy brushed surface and is produced from wool fibres that are matted together by application of heat, pressure and steam. Boiled wool although similar to felt involves a different production technique of boiling and shrinking wool by 25 to 30 times.

Duchesse satin, taffeta, silk dupion, slub silk are considered to be lustrous dress fabrics owing to their excellent drape ability and pleating properties. Accordingly, the fabrics are extensively used for designing evening bags. The bags designed with aforesaid materials when reinforced at the back are referred to as laminated.

4.3.4 Handbag construction

The construction method employed depends largely on the chosen material and handbag styles. However, the initial designing process is mandatory irrespective of the handbag styles. Handbag designers are involved in lot of brain storming exploring various sources for design inspirations. Market survey and keen observation of street fashion along with close eye on trending apparel lines prompt the designers to incorporate such designs and technical features in their initial sketch so that the end result is a quality product conforming to consumers aspirations.

The hand drawn or computer aided sketch prepared by designer usually includes all the construction, style and design details required for handbag assembling and completion. The approval of designer's sketch is succeeded by prototype model prepared in muslin or felt completed with attachment of closures and decorative elements exactly as they are included in designer's sketch.

The prototype may be modified by incorporating any changes suggested by the brand, buyer or customer and eventually taken into bulk production.

The next step includes pattern-making for all individual components of handbags in accordance with size and style specifications. Leather handbags demand fine workmanship to avoid any scars and marking and thus the patterns from one or at the most two leather layers, are cut individually using die cutting technique.

The patterns for handbags composed of textile base are usually cut from lay of several fabric plies much like ready to wear garment construction. The method ensures economy in production and saves time.

The cut handbag components are thereby assembled by machine or manually.

Three main constructions namely turned, raw edge and butted are generally used for handbag development which can be used individually or in conjunction with one another.

Turned construction involves sewing the components on the right side together and then turning inside out. Thereafter, the seams are opened and top stitched or glued down. Accurate turning requires a flexible and supple material for handbag construction.

Raw edge construction involves stitching leather pieces on the right side and leaving the material with raw edges. In this construction, the bag is not turned inside out post stitching. Vegetable tanned leather is used for raw construction due to flexibility in edge finishing by waxing, dyeing or polishing before or after handbag construction.

Butted construction on handbags involves turning over edges and glueing onto reinforcement or skived and turned onto itself before stitching together.

The assembled handbags subsequently receive stiffening materials for support and reinforcement between the body of the bag and the lining.

The shape of the handbag is retained by stays and plastic strips. Eventually, fillers made of foam are added to handbag providing softer feel. The handbag body is then fitted to the frame followed by attachment of handles, fasteners, closures and decorative embellishment.

Inspection of the final finished product is of utmost importance to check for any visible defects, abnormalities in materials, closures, etc.

Post inspection, the handbags, individually boxed or bulk packed are shipped to retail and brand outlets.

The price, quality and handbag materials dictate the mode of packing and shipment of the handbags.

4.3.5 Handbag styles

Handbag styles have crossed over genders with some bags designed as unisex entities to be relished by both genders alike. Functional bags like backpacks, tote, messenger bags and laptop bags are worn or carried similarly irrespective of gender. Today's handbags feature variety as well as functionality. The slavery of users to technology with ever increasing dependence on the electronic gadgets requires handbags that are capable of accommodating these needs. Accordingly, handbags are designed with extra pockets and compartments to house consumers' electronic gadgets. Holders and cases are specifically designed into handbag to securely carry the valuables.

The handbags are classified into varying styles based on handle styles and materials used for their construction.

Box and half-moon handbags, cell phone bag, channel signature bag, framed clutch handbag, drawstring, feedbag or bucket, hobo or slouch handbag, minaudiere handbag, satchel, tote and top flap handbags are some popular handbag styles.

Clutch is a handbag style devoid of any grasping element or handle. The style is handheld or clutched by the users. A long metallic chain or narrow leather concealed strap to carry clutch on shoulder is provided by some handbag brands.

Drawstring is a casual handbag style generally preferred by users during leisure activities. Accordingly, the bag style is designed with minimal of trimmings and fasteners and merely features a drawstring as the closure. Louis Vuitton Noe bag is a popular drawstring bag style among fashion connoisseurs.

Envelope is characterized by flat, rectangular outer body constructed in leather, vinyl or any other durable material. The salient feature of the bag is envelope shaped flat closure equipped with magnetic lock, snap rivets, etc. Envelope just like clutch may be handheld or can be designed with a concealed handle or strap to be carried on shoulder.

Box is a square or rectangular shaped, sturdy, durable handbag composed of metal, lucite or wood.

Duffle also referred to as feedbag closely resembles horse's feedbag and is generally preferred for casual applications like leisure travelling, biking, trekking, etc. The bag has substantial capacity and suitable to stuff in all the essentials that may be required by the user while on the go.

Satchel's design is inspired from doctor's medical bag. The bag features flared, flat, wide bottom accentuated with handles. The bag is generally carried on arm rather than shoulder and is popular as women's everyday bag.

Shoulder bag is carried on the shoulder that features a long strap generally made of same material as the outer covering of the bag.

Tote is designed as a shopping bag, simple in construction and devoid of any closures. The bag features an open top and two strap handles for easy handling. Totes are generally made from canvas and are used for grocery shopping or carrying books.

North-south bag is a handbag style that is taller rather than being wider and is comparable to portrait orientation.

East-west bag, unlike the former, is wider rather than taller and is comparable to landscape orientation.

Fold over bag is a flat zippered top bag that can be folded over or doubled up and grasped in hand or carried under the arm.

Hobo is a women's handbag style with a zipped top, curved crescent shape and slouchy look.

Brief case is flat, rectangular handled leather case that is usually used for carrying documents, books, files and personal items like wallets and electronics.

Cross-body bag is a small handbag with a long strap intended to be carried over one shoulder across the chest.

Fanny pack is classified as travelling bag for carrying small sized valuables. It is designed as a small-zippered pouch suspended around the waist by means of belt.

Messenger bag is a rectangular, large sized bag featuring an adjustable long shoulder strap and a single flap closure.

Suitcase is a large rectangular traveler bag preferred for carrying luggage on trips. The bag comprises solid structure and reinforced corners.

Pouch is a small, soft fabric, leather, suede bag that is gathered with a drawstring. The latter also serves as an extended handle enabling pouch to be carried along the wrist.

Travel related bags comprise the styles like backpack, duffle bag, fanny pack, messenger bag, suitcase and travel bag.

Backpack is characterized by two straps going over the shoulders and available in a variety of shapes and sizes.

Duffle bag originated as the bag catering to armed forces, however, the bag style is quite popular among civilians as well. The bag is characterized by its tubular or cylindrical shape and two straps. It uses canvas or leather in the outer body with zipped opening on top and generally preferred for carrying personal belongings while on short trips.

Travel bags are designed to cater to consumers who frequently travel by air. The bags are designed to fit well in overhead compartment of an airplane. The salient features of the bag include its rectangular shape with multitude of compartments and pockets enabling the users to carry all their belongings in the same bag.

Barrel bag resembles in its shape to duffle bag, however, is a smaller version of luggage bag.

Bowling bag is a fashionable, rigid bag that has wide bottom with an arched opening at the top and two rigid straps.

Toaster bag is similar to bowling bag. The bag features a flat bottom and curved top and two rigid straps for hanging it on forearm or shoulder.

Small leather goods (SLG) are generally carried by women in their handbags and in the pockets by men for holding small items like coins, money, cosmetics, etc. Small leather goods unlike their name are not necessarily constructed from leather but can be composed of suede and other synthetic materials simulating leather.

The category of small leather goods includes entities like check book cases, coin purses, cosmetic bags, credit card holders, eyeglass cases, key holders, cell phone cases, wallets and wristlets.

Check book case is used to hold and protect checkbook. The case is designed with slots or pockets for credit card, ID, pen attachment and can be a part of a wallet.

Coin purse is used to carry bills, coins, cards and is a small, one compartment purse.

Cosmetic bag is designed for women to carry cosmetics. It is a small-zipped bag lined in plastic on the inside.

Wristlet is a small size, rectangular, flat zipped bag equipped with short strap to fit around the wrist.

Wallet is used for carrying money, credit cards, coins and other personal items. It is a small flat bag with multiple compartments and zipped coin section.

Eyeglass case is designed to carry sunglasses, reading glasses carefully into pocket or bag. The case resembles and closely fit the eyewear.

4.4 Belts

4.4.1 Introduction

Belts are no longer considered functional accessories but have aesthetic dimension associated to them as well. Belt manufacturers cater to consumers' demands by creating their own designs to augment apparel styles, dresses, sportswear, ensembles and selling to the retailers and manufacturers.

Fabric, leather, etc. form the body of belt while fasteners like velcro, snaps, buckles, eyelets, metallic chain links and decorative embellishments are other essential components of latest fashionable yet functional belts. The following section discusses the materials generally used for belt construction.

4.4.2 Belt materials & construction

A myriad of materials namely leather, vinyl, metals, elastic, fabrics like denim, canvas and narrow woven fabrics like braids, ribbons and laces are utilized for construction of belts catering to functional or aesthetic demands of consumers. The construction method employed for belts depend on the chosen materials and belt styles.

Leather belts demand high level of accuracy and precision during their construction. Dressy and formal belts use leather as the major raw material in conjunction with adornments like nail heads, stones, snaps and rivets, eyelets, etc. Novelty belts are generally constructed from fabric, lucite, metal, vinyl, etc.

Accordingly, hand or strap cutting machines are deployed to cut the leather pieces to required dimensions considering the belt's length and width.

A traditional machine making straight cut is not suitable for shaped constructions like a contour model and thus requires its replacement with die cutting technique that involve dies to be made to match the contour exactly. Furthermore, utmost care is required to avoid any blemishes when working with leather for belt construction.

Some belt styles require backing for additional support, reinforcement and are thus backed by sewing or gluing of two pieces by aid of walking foot.

Textile materials offer flexibility and ease in belt construction owing to their ability to be stacked together as several plies in lay and cutting all the patterns together in a single strike of cutting tool unlike leather or non-textile materials thus resulting in economy in production.

Decorative embellishments by addition of flowers, ribbons or top stitching are carried out on the belts once the body of the belt has been completed. Eyeleting and buckle insertion is accomplished depending on belt style.

The construction method employed for cinch (a fabric belt tightly conforming to body by means of elasticized materials) is relatively simpler compared to their leather counterparts. The construction of cinch belt involves cutting of the elasticized materials, sewing along edges and attachment of fasteners like velcro, snap, etc for closure.

Chain link belts are metallic belts composed of numerous interlocking metallic pieces and requires fastening one link to another as per required length.

The construction of another fabric belt namely sash is faster and easier compared to leather belts as the fabric used for belt's body is just sewn at the edges and the style does not require any closure, fastener to be attached.

4.4.3 Belt styles

Cinch is composed of an elasticized material like stretchable knitted fabric, spandex, lycra in its body. Cinch is designed to accentuate the waistline, provide stretch ability and ease of wearing and taking off. The style is completed with fasteners for closure and any additional trimmings for surface ornamentation.

Contour is a belt style that is shaped specifically to fit the contour of figure and thus render illusion of slender, thinner waist.

Sash is the simplest of all belt styles with merely a fabric of contrasting color to bodice or overall attire draped around the waist.

Cummerbund is a fabric belt similar to sash that is wide in front and narrower in the back. The front of the belt is often embellished with decorative fasteners, closures, studs, artificial stones, etc.

Link is a metallic belt formed by interlocking of chain or metallic pieces. The inclusion of elastic makes the belt style expandable.

Rope is generally hand-crafted belt style that is produced using techniques like macramé, cording, braiding and plaiting that involves knotting, diagonal intersection of varying yarn strands.

Self is the term coined for belt style that is composed of same fabric/ raw material as the garment over which it is worn.

Figure 4.16 Handbag styles

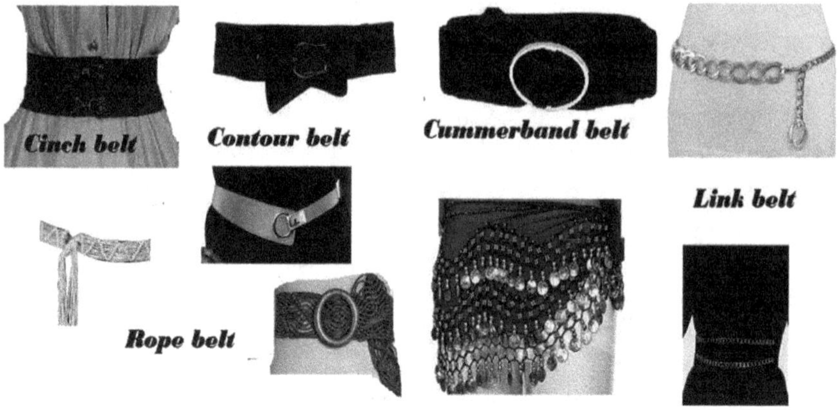

Figure 4.17 Belt styles

4.5 Gloves

4.5.1 Introduction

The contribution of fashion accessories to business and profit making to their creators and merchandisers cannot be undermined. Thus, emphasis is laid on front and forward visual merchandising of fashion accessories. The growing accessory industry prompts designers to enter into licensing agreements or affix their labels on accessories like gloves, scarves, hats and carrying knitted accessories to satisfy and content brand, label conscious consumers. The significance of gloves among both genders can be witnessed from the past times when gloves were frequently worn to royal courts, ceremonies and various occasions apart from serving as protective wear against extremes of climate.

The emergence and growth of glove industry in recent times has been tremendous with functionality and surface ornamentation both gaining consumers' attention. Gloves can be fashion focal point of females' ensemble particularly in western culture and royal families where gloves work wonders creating an enriched look for simple to expensive attire. However, weather generally dictates the usage and thereby designing of gloves to cater to the needs of consumers in cold, dry winter or warmer spring and summer season. Thus, gloves can be considered as both functional entity and a fashion statement depending on wearer's preferences. A white sheer lace mid arm long gloves adorned by brides in western culture complement the bridal attire well.

Nevertheless, driving gloves, protective gloves and winter wear gloves impart a functional dimension to this accessory category.

Gloves in recent times are considered functional accessory rather than an embodiment of status or social rank. The gloves initially enjoyed the status of being merely a hand covering comprising of an animal skin. However, the evolution of highly specialized gloves for specific occupation like protective gloves for healthcare, paramedical, firefighters, field workers and astronauts has been a remarkable transition. Gloves have witnessed dramatic transformations right from long cuffed ones with appliques and embroidered details to cut resistant, breathable workwear gloves.

4.5.2 Glove materials

The raw material selection for the gloves is based on their price point, style and end usage. Dressy gloves are designed for high-end, elite consumers who generally prefer patent leather for their accessories. Accordingly, supple durable, pliable, expensive leather like cowhide, lambskin, pigskin along with cabretta, capeskin, deerskin, buckskin, doeskin, kidskin and goatskin are preferred for construction of dress gloves.

Work gloves worn for health, hygiene or protection against environmental, occupation hazards are constructed from horsehide, suede either alone or together with cotton twill fabric. Fibres exhibiting thermal insulation property namely wool, acrylic, mohair, cashmere are preferred for winter wear gloves. Enhanced protection against wind, cold, moisture is ensured by using high performance, breathable materials like Teflon, Gore-tex. The materials as coating or membrane can be applied to textile substrate.

The insulation properties can further be improved by inclusion of high-performance materials like Thinsulate by 3M and Polartec in gloves intended for extreme cold weather conditions. Warmth can further be infused in winter gloves by fiberfill or down feather sand witched between lining and outer fabric of gloves. Lining material for gloves vary from flannel, fleece, velvet to fake and real fur.

Gloves embellished with gold and other precious metals and stones were adornments for rich, effluent class as an indication of one's status. Furthermore, effluent, upmarket consumers would seek a matching pair of gloves with other accessories like footwear and belts to flaunt their fashion sense. Nevertheless, functional aspect of the gloves is at forefront with glove designers focusing on designing gloves for cold weather and for specific purpose like protection of hands against chemicals, bio-hazards, fire, etc. The following section discusses the various raw materials utilized for glove construction in details.

Deer, goat, lamb and sheep leather is widely used for making high-priced gloves. Exotic skins are characterized by unique texture, grainy effect and aesthetic appeal. The skins are obtained from ostrich, crocodile, python, lizard, mink, fox. Exotic skins are preferred for design details and cuff trims as the extensive usage in the entire glove body would raise the price considerably.

Leather used for glove making should be checked for suppleness, fineness, softness and elasticity during tanning. The most suitable and preferred leather for glove designs are the ones that are strong, featuring fine, even grain. Quality and stretch of leather determines the fit of the glove. The patterns for gloves should be cut precisely so that stretch goes around the hand and not stretch running the length of hand and fingers. The latter method of cutting would result in unnecessary stretching of gloves each time the wearer pulled the fingers to remove the gloves.

Cabretta, kid skin, deer skin, peccary, slink lambskin, slink calfskin, sheepskin, goatskin, cowhide and exotic skins are some varieties of leather generally used for exotic, high-end glove making.

Carbetta leather - features a smooth, regular and refined surface, softness and even grain. Strength and suppleness of this hair sheep leather makes it comfortable for next to skin applications. Another advantage that carbetta offers is varying texture that can be imparted by finishing it by glacing, nubucking, sueding or lacquering.

Kidskin - has a finer grain than goat skin and is popular for its velvet soft texture.

Deerskin - softness, suppleness along with exceptional strength and elasticity makes this leather a favorable choice for glove designers.

Peccary - the source of peccary skin is a wild pig inhabiting in amazon jungle. The leather has unique, characteristic cluster of three dots over its surface, and is known for its strength and softness.

Slink lambskin - is characterized by its incredible softness, suppleness and exorbitant price range. The source of skin is premature lamb.

Slink calfskin - is soft leather and highly valuable for glove making. The source of leather is still born calves.

Shearling/ sheepskin - has a unique furry lining and thus find application in making winter gloves and mittens.

Goatskin is hard wearing and coarse leather although it lacks the handle and stretch ability required for dress gloves. Thus, the leather is generally used for making gloves in lower price points.

Cowhide is a thicker and bulkier alternative compared to other leather varieties used for dressy gloves. Thus, it is preferred for casual, cheaper gloves and not an ideal choice for dress gloves.

Stretch fabrics like lycra, elastane, spandex, etc. and stretchable fibres blended with silk, velvet, nylon are used for formal evening gloves. Stretch crochet net, stretch PVC and stretch duchess satin are other options for casual as well as formal, dress gloves.

Polyester, fleece, nylon nets, vinyl, cowhide, goretex, Teflon coated fabrics are preferred for designing sports glove,

Silk, leather have been traditionally used for glove making however, the incorporation of specialized properties like waterproofing, breathability, etc. requires high performance, smart materials for glove construction. Accordingly, breathable materials like Gore-Tex have been used for designing sportswear gloves. Gloves made of Gore-Tex provide comfort to wearer by enabling sweat to escape out but do not allow liquid water to penetrate inside. Stitched seams in the gloves are being replaced by seam sealing in protective gloves. Waterproofing in the gloves along the seams can be achieved by treating the stitch holes and seams in gloves composed of waterproof fabric to render them resistant to rain or snow.

Bike gloves are designed using high-performance materials like Carbon, Teflon and Kevlar to impart protection to wearer's hand while driving. Carbon fibre form the knuckle part of the glove, Kevlar imparts protection to glove against cuts and Teflon coating ensures extra protection. Combination of toughness and protection with sensitivity and flexibility is the key issue to be addressed while designing gloves to be worn by astronauts along with spacesuits and pressure suits in outer space.

Cut resistant gloves specifically designed for work environment prone of laceration ensure maximum protection for workers. Hand and finger injuries are considerable during industrial accidents and thus hand protection is of paramount importance for occupational hazardous conditions.

High performance, specialty fibres like glass fiber, Kevlar, HPPE, carbon fiber are preferred for cut resistant gloves. HPPE and aramid fibres on account of their high tenacity, modulus and good abrasion resistance enable them to be used for construction of seamless gloves.

Low moisture regain of HPPE makes the glove resistant to bacterial growth and can be readily sanitized too. Medical glove liners are used for protective hand wear for health care professionals. The gloves used by health care professionals should isolate the hands from external biohazards and provide protection to hands from accidental nicks and cuts.

HPPE fiber glove liners are generally used for such applications owing to high level of cut protection and tactile sensitivity. However, the knitted structure of HPPE fiber glove liner fails to provide protection from accidental needle piercing into skin.

Glove liners outshine latex glove for medical professionals as the former can be sterilized. Gloves intended for chemical protection utilizes natural rubber, butyl rubber, neoprene, nitrile, polyvinyl alcohol, PVC, polyethylene, Teflon, PTFE, polytetrafluoroethylene and viton.

Chemical protective gloves should serve as protective barrier against toxic chemicals. The materials used for designing gloves against toxic chemicals vary widely based on their ability to resist permeation and penetration of chemicals.

Fleece, acrylic, wool and polyester are the commodity fibres generally preferred for winter gloves while latex, vinyl, rubber, etc. are used for functional, workwear gloves.

Medical gloves are the safety, protective gloves for patients and health professionals against exposure to infections, microbes or bodily fluids. The gloves are designed to be close fitting and generally made of latex. Additionally, vinyl, nitrile rubber, etc. can also be used to prevent allergy.

Latex and rubber gloves are generally used to protect the hands against soil, dust, germs and household routine activities like washing, cleaning, gardening, etc.

Chainmail gloves also referred to as armor gloves are composed of small metal rings linked in mesh pattern. Woodcutters, butchers and scuba divers generally wear these gloves.

Lining fabrics should not add any additional bulk to gloves and hence silk, polyester, finely knitted acrylic and wool are preferred as lining materials for gloves. Some other lining materials include flannel, fleece, fine grade cashmere knit, cotton knit and woven fabric, etc. and are generally chosen based on required bulk, flexibility, warmth and level of protection required in gloves.

4.5.3 Glove components

Gloves can be composed of as many as ten parts based on the style and method chosen for their construction. The closer fit is ensured by designing gloves with more pieces or components. The exception to this concept is gloves using stretch fibres like spandex which create closer fit with minimal glove components. Knitted gloves are constructed as a single piece owing to inherent stretch ability and conformity of knit structure as per dimensions of wearer's hand.

Leather, woven gloves are composed of up to four individual components to impart a snug, proper fit and comfort to the wearer.

The main components of gloves are as follows:

Trank is a rectangular silhouette or shape constituting the glove's front and back portions.

Fourchettes are referred to as fingers in French. Extra space and fitted shape to the fingers of gloves is provided by narrow oblong strips or pieces. Fourchettes are also referred to as gusset or sidewall. Trank and fourchette can be made in the same material or as per the design and comfort aspects, the material and color for the two can vary. Knitted fourchettes as inserts are generally used for sporty gloves.

Quirks are the small triangular or irregular diamond shaped sections fitted at base of fingers and thumb added to provide expansion and flexibility to wearer's fingers. Additional comfort, flexibility and contoured fit is ensured by inclusion of quirk in the gloves.

The most exclusive and expensive gloves are composed of double fourchette and quirks. The portion that serves as a gusset fitting inside two adjacent fingers is referred to as fourchette while quirk connects fingers at the base of the fingers.

Thumb is a separate component that is attached to the hole cut in the trank.

There are different thumb types based on how they are attached to trank

English or Bolton thumb is a bulky thumb in which thumb quirk and trank are cut in one piece. The thumb in glove provides easy thumb movement and generally used in men's dress and sport gloves and female casual gloves.

Quirk thumb also referred to as French thumb is a two-piece component, generally used for women's dressy gloves owing to its snug fit. Thumb and thumb quirk are stitched together followed by inserting into hole cut in the trank.

Set-in thumb is sewn into an oval hole cut into the trank.

Novelty thumb extends to the cuff.

Cuff protects the wearer's wrist or forearm as it extends beyond the palm. The knitted cuff provides a close fit and features a slit for easy removal.

Lining is the functional component to cover interiors of glove and provide support. The lining should be secured to the interiors of the glove so that it doesn't invert as the gloves are taken off the hands.

The material for lining may vary as per the application area of gloves. Fur and wool lining is preferred for winter wear gloves while sleek, dressier gloves are lined with silk, rayon, etc. However, knitted gloves are generally unlined where the lining is altogether absent.

Full lining may be used to cover the entire glove. Alternatively, skeleton lining which just lines the trank portion of glove may be used in some glove styles.

Domes are the metallic fasteners (snap) for fastening at the wrist.

Hearts or stays are the reinforcing sections on the glove's palm or trank portion.

Bindings are tapes attached to wrist of the gloves for edge neatening or finishing the edges.

Gore flare on the cuff is provided by a triangle shaped inset of cuff referred to as gore.

Pointing also referred to as silking is the application of three rows of decorative stitching on the glove's back portion.

4.5.4 Glove construction

Construction of gloves involves consideration of basic requirements of dexterity, comfort and the construction methods can vary as per the styles, materials and usage of gloves.

Work and dress glove manufacturing encompasses traditional and labor-intensive production process. However, active sportswear gloves are manufactured with advanced technologies incorporating high-performance fibres and finish application to achieve low bulk and higher insulation in gloves while fashion is the focal point for novelty knitted glove manufacturers.

Designers experiment with a range of colors, materials and detailing while designing different glove styles. Inclusion of bright colors, floral and contemporary patterns, varying fastening elements and trimmings enhance the aesthetic appeal of gloves.

Matching parent-child set, bridal gloves embellished with sheer trimmings, ribbons, etc., specialty gloves like gauntlets for gardening and workwear gloves, protective gloves are all designed considering the prospect user and their specific requirement of wearing gloves.

Fully fashioned knitted gloves are one size fit all gloves owing to the stretch ability of knitted fabrics. The knitted gloves thus require minimal hand workmanship as they are fully automated. However, snug fit demands precise cutting and sewing of leather and woven gloves.

Leather glove making process involves processes like taxing, table cutting, sewing and pointing. *Taxing* involves determination of exact pairs of gloves that can be cut from selected leather layer by careful examination and marking of the skins. The skins should be preferably dampened and stretched to obtain proper fit of the developed gloves.

Table cutting is generally employed for top grain and finer leather qualities. The process involves dampening the leather, pulling it over table's edge and determination of the correct shape before the trank of the gloves can be cut manually. Cutting accomplished by cutter must ensure maximum flexibility for the glove's wearer.

Pull down cutting can be used for split, slab and other less expensive leather types. The procedure unlike table cutting involves minimal manual intervention and does not necessitate the careful manipulation of each pattern of leather. Pull down cutting involves the use of die cutting technique to obtain the different patterns of gloves like trank, fourchettes, thumb, etc.

The trank cutting is followed by lay formation wherein leather plies are piled one above the other to cut the fingers and creation of thumb holes by steel dies.

Two types of patterns namely clute cut and gunn cut are used for leather glove production.

Clute cut pattern cutting encompasses cutting a full one-piece palm shape along with fingers (that are cut along with palm). As the fingers and palm are cut together as single entity; the glove style does not require any seaming across the palm and ensures comfort for the wearer.

Gunn cut process involves cutting fourchettes or fingers separately from the palm and is more suitable for gloves constructed from heavier leather.

Clicking The process of cutting the individual glove patterns by specially designed metallic dies is referred to as clicking. Sharp dies that have the ability to cut through several fabric or leather layers are used during clicking. Steel die cutter resembles four finger shaped cookie cutter. The dies are clamped in hydraulic press which ensures smooth cutting of patterns through the layers of chosen materials.

Assembling techniques Different assembling techniques are employed to assemble the different components of gloves. The chosen assembling technique depends on material chosen for glove making, its style and price point.

Inseaming, outseaming, overseaming, pique and half pique seaming techniques are used for glove construction.

Inseam technique is the least expensive assembling technique whereby no visible seaming is used. The seams are sewn with back and front pieces with right side together.

Outseam is opposite to inseaming and creates an exposed seam on the face and the raw edge of the glove. Stitching is done along the material edge. The exposed stitching in outseam can be protected by stitching a leather covering known as davey tip onto it.

Overseaming resembles outseaming however, the former uses overlock stitch to conceal the exposed seam and impart a casual look to the glove.

Pique seaming technique is preferred for comfort and sleek fit in the glove. The seaming technique requires specialized machinery for its construction and is thus one of the most expensive seaming operations for gloves. One edge of material is sewn over the other on front as well as back exposing just one raw edge.

Half pique technique - variation of pique seaming technique is a less cost intensive half pique process. This technique involves overlapping of seams on the back while other portions (front) of gloves use less expensive inseam or outseam closure.

Pointing - Some glove constructions are decorated by pointing or silking that involves application of three rows of top stitching on back of the glove.

The assembling of the gloves is followed by placing the constructed gloves on forms for size and fit conformity and then pressed.

The construction process for woven and leather gloves is similar except for the table cutting followed for leather. The woven fabrics owing to uniformity in terms of appearance, quality and visible grain lines do not require as much precision in cutting as with fine quality leather.

4.5.5 Glove Styles & glove length

Designers most often include the gloves in fall/winter lines and attempt incorporating features that match with technology driven lifestyle of consumers. More recent trends prevalent in gloves is fingerless glove just covering the lower knuckles of the hand. The design features enable the easy operation and handling of touch sensitive screens of smartphones. Design aesthetic of the line and target consumer dictates the styles of gloves.

Stylization of gloves is based on the raw material, decorative trimmings and exposure/ coverage of wearer's hand. A few popular glove styles are being discussed in the following section:

Mittens are designed keeping in view children who find it difficult to put on and take off the gloves and are not very skilled in so doing. Mittens with only thumb separated from rest of the fingers are an easy option for kids as they can simply insert the thumb separately.

Shorty as the name suggest, is short length glove – a two button glove reaching up to wearer's wrist. Leather or woven shorty may be provided with a side or centre opening for ease of wearing. However, shorty gloves composed of knitted fabric do not require wrist opening owing to inherent stretch ability of knitted fabrics.

Biarritz Slip on glove is devoid of any fastener and slit opening can be slipped on easily. Slip on glove can be two button ending at the wrist or four – six button reaching up to the arm.

Gauntlet features a triangular insert at the wrist and flared cuff.

Mousquetaire glove style varies from eight to sixteen buttons in length and generally complements women's evening wear attires. The glove features a vertical opening above the wrist enabling the easy doffing and donning of glove by wearer through fingers and without having to remove it completely.

Driving gloves closely resemble the gloves of formula car drivers and features perforations or ventilation holes so as to provide easy sweat absorption from wearer's palm. Furthermore, the design of the gloves ensures comfort and flexibility to wearer. Soft leather, suede, woven fabric along with knitted fabric used in trank of gloves enables a firm grip of steering wheel to wearer while driving.

Fingerless or waif glove exposes the fingertips as the fourchettes of the glove end just below center joint of hand. Fingerless gloves are popular among sports freaks and gym enthusiasts.

Work gloves are worn for protection against external hazardous elements like microbes, chemicals, fire and molten metal splashes. The gloves find application in health care, hygiene and industrial applications. High-performance fibres, fabrics, latex and leather are generally used for the work wear gloves.

Glove lengths

Gloves are designed in varied lengths ranging from those that end at wrist to the ones which reach up to the mid arm and the shoulder. Button is the term used to express the length of gloves where 1 button corresponds to one inch, starting from thumb's base. Accordingly, one button glove is the shortest that reaches to the wrist while two button glove is longer and covers the wrist. Four button glove is designed to reach between end of wrist and middle of arm. Eight and twelve button gloves are longer with the former ending at the middle arm and the latter at the elbow. The longest glove is a sixteen button glove that reaches up to the shoulder.

4.6 Knitted accessories

4.6.1 Introduction

Functionality along with fashion is the focal point for designers while designing knitted accessories like socks, stockings, stoles, mittens, gloves, etc.

Hosiery items like bacteria-controlling athletic socks may be designed with sole objective of providing comfort to wearer while fishnet hosiery, pantyhose may just be designed as subordinate and complimentary element for apparels. The production of knitted accessories is predominantly an automated process that involves very little hand labor. The manufacturing can flourish at domestic levels due to its non labor-intensive nature.

The materials chosen for knitted accessories depend on the intended end-use. A variety of fibres namely cotton, acrylic, polyester, wool, nylon, spandex and silk are utilized for knitted accessories. Retro fashion trends go a long way in inspiring the designers for their creations. Designers experiment with fashion statement by offering patterned, opaque, sheer, fishnet hosiery as leggings, jeggings, track pants and pantyhose targeting female consumers.

The inception of concept of "Novelty Wellness" changed the outlook of designers and manufacturers towards designing of knitted accessories. The concept has revolutionized the knitted accessory industry with prime focus on providing age defying, rejuvenating, anti-cellulite, body enhancing and energizing characteristics to the prospect, health-conscious consumers.

Anti-stress pantyhose treated with plant extracts, fatigue free knitwear enriched with topical healing sustainable agents like aloe, gingko, biloba, ginseng, green tea are some of the path breaking innovative concepts employed by designers considering the well-being aspects of consumers. Hosiery items like tights are being designed with incorporation of leg softening aloe Vera extract, rosemary, apricot moisturizers in the form of tiny, scented moisturizing beads. The moisturizing beads get released due to friction of the tights against legs during motion. Some other innovations in knitwear accessories include thong toed sheers, toeless sheers and non-slip pantyhose. The non-slip sheers are best to pair with sling back pumps, stilettoes as they provide traction for the footwear. Power socks by Adidas are an innovation in men's sock segment. The socks designed as a tight fitted, over the calf accessory is capable of compressing the leg muscles particularly during running. Muscle oscillations are reduced and circulation enhanced as the wearer is involved in any dynamic activity due to graduated compression thereby increasing the runner's stamina. Graduated compression hosiery boasts of enhanced performance for sportsperson and also targets ageing women. Legg's sheer energy hosiery follows the league of concentrating on wearer's comfort, however, it targets broader spectrum of women consumers.

Saturation in the existing knit accessory market has led the designers and retailers to quest for novelty and innovations in knitwear styles. A day is not far ahead when designers will come up with depilatory features in leggings and pantyhose possessing capabilities of providing wearer the feel of smoothly shaven legs without using razors.

4.6. 2 Materials & construction methods for knitted accessories

A wide range of materials are used considering the feel, comfort and durability aspects desirous by consumers in knitwear/knitted accessories. Accordingly, nylon, micro fibre, cotton, silk, wool, rayon, polyamide, acrylic, polyester and spandex are some of the commonly used fibres preferred by designers for designing knitted accessories. Nylon microfiber is preferred when sheer look along with lightweight, durability and comfort are required in knitted accessories. Absorbent fibres like cotton, rayon are used in panty hoses, tights for breathability and suitability in summers. Commodity fibres and their blends like cotton, silk, wool, rayon, and polyamide are widely used for making socks.

High performance, specialty fibres are becoming designers and manufacturers' favorable choices owing to innumerous advantages associated with the fibres and the products designed thereof.

Organic cotton, wool, silk, antibacterial and wicking fibres are becoming popular among designers of knitted accessories for creating sports socks.

The use of wicking fibres in socks enable dry feel next to wearer's skin owing to rapid sweat absorption and dissipation by these fibres. Apart from keeping moisture away, hollow and tetra channel, trilobal polyester fibres possesses higher thermal insulation thus providing warmer feeling to wearers.

Knitted accessories primarily focus on creating look of natural luxury fibres with an ability of product to be well fit and machine washable. Cashmere, mohair, merino wool with or without finish can also be used for designing high-end knitted accessories ranging from stoles, scarves, mittens to beanies and socks.

Metallic threads like lurex, texturized yarns and ornamental elements like applique are extensively utilized to cater to teen and twine market designing evening legwear, decorative, lightweight, warm stoles.

Stockings, pantyhose, knee high stockings and bobby socks are the commonly used knitted accessories. Stockings serve as leg coverings intended to be worn with casual attires like dresses, skirts, etc.

The stockings usually are made as seamless structures on circular knitting machine or can also be knitted on flatbed knitting machine and sewn by application of seams. Plain stockings or those featuring embellishment, embroidery on ankle are popular among females. Garters are worn along with stockings to prevent slippage.

Seamless knitted accessories are produced on circular knitting machines. Foot, leg shape for legwear and trank, fourchette for gloves can be created by fully fashioned knitwear technology. The technology can produce the required shapes simply by dropping or addition of knit stitches on the knitting machine. The shape of leg or hand is imparted to knitwear by permanent heat setting operation referred to as boarding. It involves pulling the designed knitwear on a metal form thereby producing the accessories resembling the shape of leg or the hand.

Pantyhose is a one-piece configuration combining stockings with a panty, thereby preventing slippage and eliminating the need to wear garters. Spandex and other elastic materials are used for construction of panty hose to provide support to the wearer and control excess bulges. Knee high stockings ending just below knee are another range of knitted accessories to be worn under pants. Socks of variable length and made in different materials are equally popular among females to be adorned with formal, casual, occasional attires. Formal hosiery caters to the men's sock segment. Men's socks are available in a range of length, materials, patterns and styles.

Fully fashioned stockings are produced in flat form as per the designated size and shape. The assemblage is finished by stitching the back seam. Circular knitting machines are employed to produce seamless, one-piece knitted accessories. The seamless pantyhose, stockings are given the desired size and shape on the machine and subsequently subjected to dyeing and heat setting. The construction of pantyhose can be accomplished in two ways. An individual stocking and panty are constructed and then stitched together followed by dyeing and heat setting. Another method of constructing panty hose involves the use of specialized machine to create it as one piece. The construction completes with dyeing and heat setting of the pantyhose. The similar construction stages are followed for socks as with stockings and pantyhose. Socks are generally knitted on circular knitting machine using colored knitting yarn to eliminate the need for dyeing later on. Else, the socks can be knitted in natural color, bleached and then later on dyed to achieve the desired shades. Nylon and one of its exotic variety Enkasheer along with fibres like cotton, wool, cashmere, acrylic, and polyester are the fibres of choice for construction of knitted accessories. Colors of knitted accessories may change from season to season and they present a glimpse of hues, shades prevalent in apparel market.

Apart from socks, stockings and pantyhose, knitted fabrics are also used for construction of gloves. Knitted gloves provide an additional advantage of stretch ability and conformity to wearer's hand. Commodity fibers like nylon, polyester, fleece, acrylic, wool, cotton, vinyl are generally used for knitted gloves.

Knitted gloves are generally unlined and are much simpler to construct compared to gloves composed of woven or non-textile materials like leather. The latter involves several construction steps like taxing, table cutting, pointing. However, knitted gloves demand little assembling as it can be knitted as a single piece on circular knitting machine. It requires minimal stitching just for closing the fingertips. Knitted gloves are basically one size fit fully fashioned gloves that are fully automated and do not require intricate, time consuming, manual workmanship like leather gloves. The simpler and fewer steps involved in the production process of knitted gloves make the process faster and economical. A need for separate sewing of gussets to be inserted between fingers can be completely eliminated for knitted gloves where addition and dropping of stitches during knitting can produce gussets along with the trank or palm portion of the glove.

4.6.3 Styles of knitted accessories

Socks and hosiery can be classified into different styles based on foot, leg coverage, materials and trimmings employed for their creation.

Footies also referred to as golf or tennis socks is the style of socks that end under the ankle and with minimum visibility at the shoe tops.

Anklets also known as socklets are socks featuring small turned down cuff that cover the ankle._

Bobby socks - this class of socks is a heavy cuffed anklet generally preferred by girls.

Crew is a ribbed, stretchable sock style that reaches mid-calf.

Hiking socks are crew style socks worn with hiking boots. These socks feature thick ribs that protect lower legs from chafing.

Over the calf is men's sock style that reaches up to the knee and ensures that skin is not revealed as men are seated.

Tube socks are knitted in tubular shape and are free of knit in hccl. Thc socks are used by males as athletic socks.

Knee highs or trouser socks are worn with slacks and generally reaches up to the knee. The socks are available in range of patterns and colors from sheer to solid colors.

Over the knee socks is a female sock style that covers the knees.

Leg-warmers - designers of this hosiery style took inspiration from dancers who used to wear leg warmers of varying lengths from ankle to thigh to prevent leg cramps.

Thigh-highs reach up to the mid-thigh and serves as an alternative to panty hose. The use of garters is eliminated owing to the presence of top lacy, elasticized band preventing the hose from slipping.

Leg-highs are similar to thigh high, the only difference being slightly longer length and use of microfiber band rather than lace band as in thigh highs.

Stockings can be credited for adaptation of most hosiery styles. Stockings are similar to thigh high but the function of garter or elastic top band here is performed by a clasp. The clasp secured to lingerie ensures that the upper stocking is held in its place.

Pantyhose are preferred knitwear styles adorned with short skirts and are composed of stockings and panty knitted as a single entity.

Tights are generally adorned by women while performing aerobics, dancing, etc. They are the opaque versions of pantyhoses.

Body stockings are laced stockings extending from bust to toe.

Demi toe socks generally worn with open backed footwear, features a reinforced toe and heel portion made of nude, sheer material.

Fishnet socks and stockings feature a unique diamond net pattern that is created by warp knitting technique.

Body cut knitwear style resembles a pair of shorts by the way panty portion drop slightly down on thigh.

French cut is worn to give illusion of longer legs as panty portion is designed to be higher on the sides of thighs.

Maternity knitwear as the name suggest is designed for pregnant women. The style characterizes an extensible front panel and compressive support to aid circulation in legs.

Opaque is knitwear style preferred in cool climate and accordingly is designed thinner than tights but heavier than sheer hosiery.

Reinforced toe or heels are provided in socks and stockings to prevent these stress prone areas against whole generation or puncturing.

Sheer toe waist stockings and pantyhose are designed with sheer fabric from toe to waistband and do not consist of reinforcement in any area.

Toe socks resemble a feet glove with individual components for each toe. These are decorative socks featuring horizontal strips.

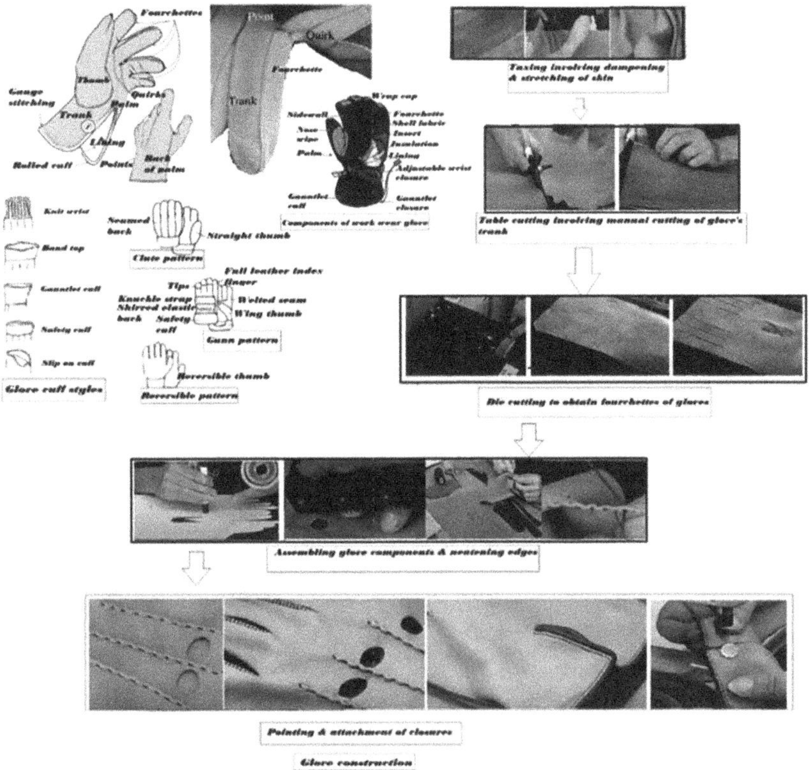

Figure 4.18 Glove components & construction

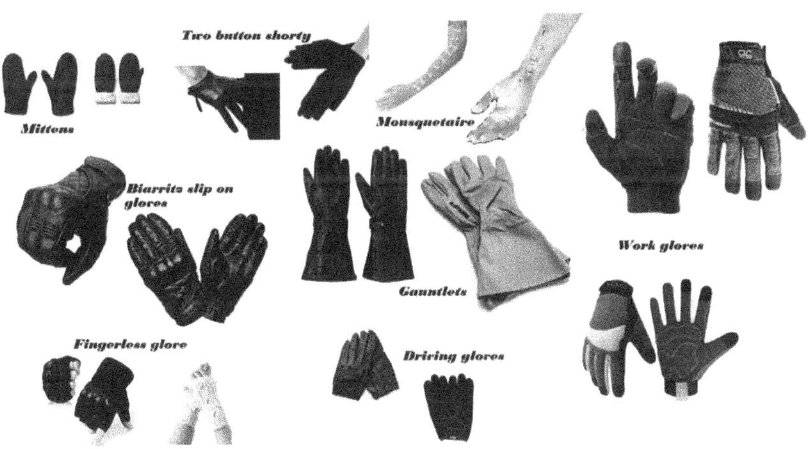

Figure 4.19 Glove components & construction

Sandal foot socks are designed to be worn with between the toe footwear, sandal like thongs where a piece of metal insert may separate a big toe from smaller four fingers. It is composed of a sheer toe and heel without any reinforcement.

Support is referred to any knitted accessory using combination of spandex with nylon and any other synthetic fiber to exercise better control.

Mesh knitwear are composed of mesh knit structures.

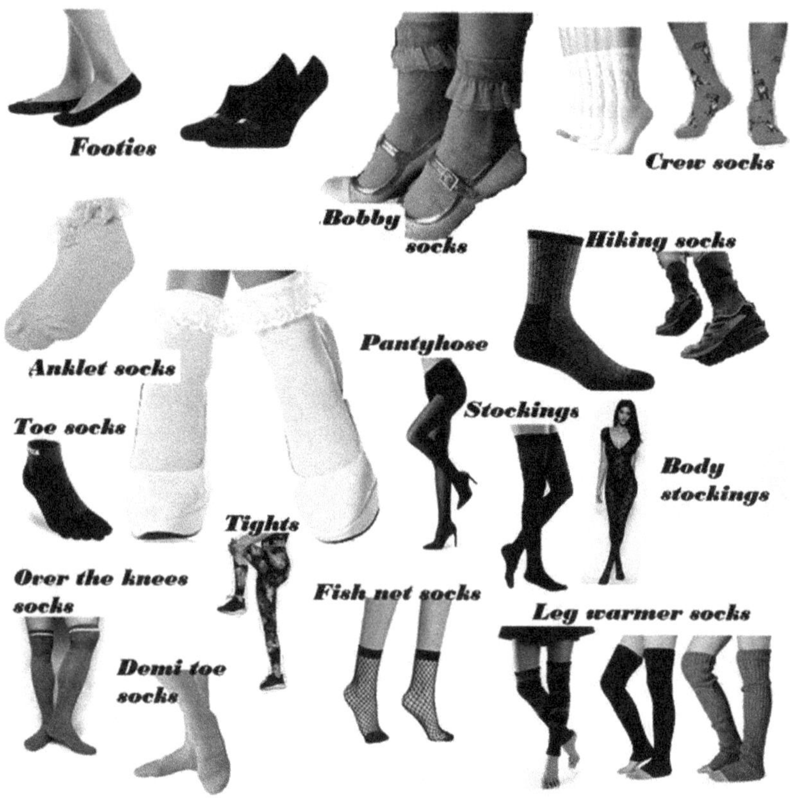

Figure 4.20 Styles of knitted accessories

4.7 Smart wearable accessories

4.7.1 Introduction

Smart or high-performance textiles refer to the textiles that can sense external stimulus and respond as per the sensed stimulus. Smart clothing is classified as passive, smart and ultra-smart based on their ability to respond and adapt themselves in accordance with external stimulus like mechanical, thermal, chemical stimulus or changing ambient conditions.

The consumers' quest for latest technologies in their clothing like attending calls, calories burnt, steps moved, GPS enabled clothing, color changing, monitoring of physiological parameters has fostered the designers and manufacturers to come up with wearable devices, sensor integrated clothing, color changing, thermal and moisture regulating, protective, light emitting, antimicrobial and high-performance clothing. Furthermore, designers and manufacturers are exploring the accessory industry in order to incorporate the smart features and enhanced functionalities in an otherwise ordinary looking accessory. The consumers may be startled to discover an ordinary looking pair of earrings serving as earphones, smart watch enabled with GPS, location sensing, or googles that can play music.

Accordingly, many accessory brands are exploring this field and attempting to blend technological aspects with aesthetics for design and development of smart wearable accessories. The following section discusses the latest innovations that accessory industry has witnessed by inclusion of smart, high-performance features in the decorative and functional accessories.

4.7.2 Innovative features in smart accessories

Planet beyond is an accessory brand that has blended aesthetics with technology in its accessory range. The earpieces designed with eye catching metallic leaf design laced with technology in first glance presents resemblance to any exquisite pair of old, silver earrings. However, the earpieces serve as earphones crafted sustainably from recycled materials.

Series 4 smartwatch of Apple features GPS and cellular functions that facilitate heart monitoring, workout patterns and can serve as a smart phone too. It showcases a color changing display.

Louis Vuitton's smartwatches unlike other smart watches editions primarily focus on travel rather than fitness and health monitoring by other smartwatches. The highlights of the smart watch include LV guide and my flight for wearer to experience bespoke luxury travel recommendations as the watch senses the location of the wearer.

Bose frames are the audio augmented reality shades equipped with built in sensors, hidden speakers and microphones. The frames appear to be like any other ordinary black nylon-framed sunglasses with polarized lenses but only wearer knows the delight of wearing the glasses listening to their favorite soundtrack and protecting eyes from sun's glare simultaneously.

Altrusis by Vanaya designs assortment of smart jewelry – necklace, rings and bracelets. The smart jewelry is capable of connecting to smartphone, sending vibration alerts and can be customized as per wearer's needs.

Bella beat leaf urban is an activity tracking pendant which can be worn as necklace, clip or bracelet. The salient features of the smart transformable pendant are sleep monitoring, stress sensibility and menstrual cycle tracking for the wearer.

Ringly luxe smart rings and bracelets serve as an activity tracker by recording steps, distance walked and calories burnt by the wearer. Furthermore, the ring reminds the wearer to indulge in meditation, breathing exercises and can provide customized alerts for phone notifications.

Michael Kors Access bracelet is equipped with a link feature that enables the wearer to take selfies, play music tracks and run presentations. Moreover, the bracelet also serves as an activity tracker laced with sleep monitoring.

Senstone clip, pendant or bracelet is distinct from other smart, activity tracking accessories in the sense that it is a voice recording device recording voice notes on the move. The smart piece of jewelry has capability of creating audio notes and can be integrated with smart phone apps.

Joule earring backings can be paired with any earrings as it is concealed behind the ear. The smart wear performs multiple functions like activity tracking and health monitoring by providing data like steps covered, calories burnt, heart rate, etc.

Fiber optic accessories namely cap and rucksack had been launched in New York fashion week. The designed cap and rucksack enables the wearer to change the surface colors and patterns with the changing music played via accompanying music app on a smartphone.

Thesis couture came up with high-performance stilettoes that incorporate ballistic grade polymer in the heel stops. The design features of the footwear such as the angle of shoe and platform at front gives the illusion of wearing a low height heel to wearer along with comfort and painless underfoot experience.

Cross trainer athletic shoes boast of the technological innovations ranging from motion control to shock absorption. The footwear are designed as per the requirements of specific sports. The athletic shoes are designed with lightweight components and provide added comfort to wearer due to breathability, cushioning effect and shock absorbency features incorporated into smart footwear. Apart from sports footwear intended for specific sports, shoes designed as athleisure are also gaining worldwide acceptance by health-conscious consumers. Casual fitness footwear are laced with technologies like in sole cushioning, molded and contoured foot beds, padding, shock dispersion, reinforced uppers and air ventilation. The shock absorbing features provide overall comfort to wearer's feet relieving them of standing and walking fatigue.

Latest technology adopted by Nike in lifestyle footwear for entrepreneurs incorporated polyurethane capsules filled with pressurized gas in footwear's insole. The comfort technology introduced by Nike promises a cushioning and comfort feel to wearer's feet just like walking shoes.

Footwear technologists have come up with orthotic features in footwear specifically designed to relieve the ageing feet of stress. Orthotic footwear is equipped with support device for feet and ankles. Further, in order to ensure even distribution of pressure on bottom of foot, impact absorbing, shock diffusion inserts are equipped in the footwear.

Everyday stress encountered on feet due to mundane physical activities by wearer can be relieved by inclusion of heel pads, cushioned insoles and arch supports.

Siren socks feature micro sensors woven into the fabric to continuously monitor wearer's foot temperature. As an elevated foot temperature (resulting in inflammation) is detected by socks, the smart accessory sends an alert via text message or notification on user's smartphone. The socks are particularly beneficial for diabetic users who suffer from nerve damage and the symptoms can remain ignored owing to numbness in wearer's legs.

Sensoria's connected socks featuring three textile pressure sensors, are capable of measuring the pressure exerted on the wearer's foot during running, can track wearer's run and provide information pertaining to pace, distance, time and style of running.

Owlet Smart Sock is based on pulse oximetry technology and finds application in hospitals for monitoring the heart rate of infants. The real time data can be obtained via interconnection of smart sock to doctor's iPhone or Android phone.

Neopenda smart baby hat constitutes of a monitor that is fitted inside a hat for infants. The hat is capable of measuring temperature, heart rate, respiratory rate and blood oxygen saturation level of infants. A custom software can be operated via a tablet that receives the information when wirelessly synced with upto twenty four such baby hats, thereby enabling doctors and nurses to monitor vital signs of the whole room at a glance and can be updated in case of criticality or variation in temperature, heart rate of babies in the ward.

Compression sleeve based on electrocardiogram (ECG) technology, can be used in hospitals for monitoring patient's heart rate activities. The sleeve apart from offering accurate heart rate data, is laced with sensor that enables monitoring sleep patterns, body temperature, stress levels workout intensity, detection of heart inflammation and coronary heart disease, air quality and UV rays.

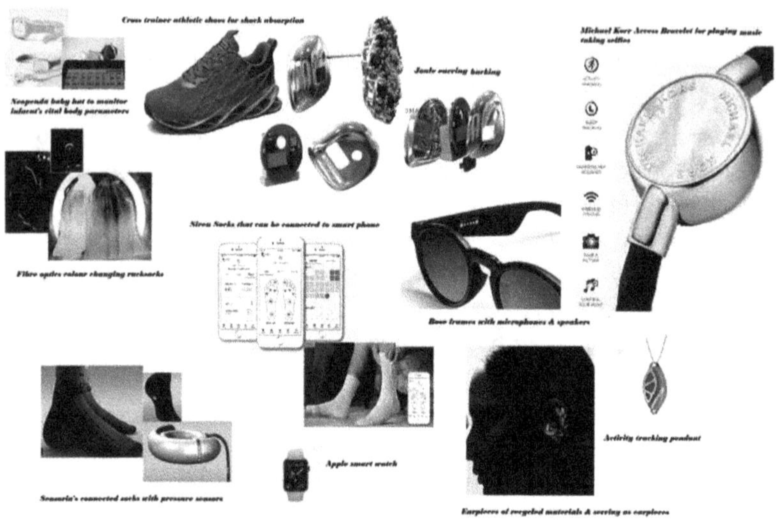

Figure 4.21 (a) Smart/ high tech accessories

Figure 4.21 (b) Smart/ high tech accessories

The Unseen for Selfridges have developed line of luxury smart accessories that are capable of changing colors in response to external ambient conditions or stimulus like air pressure, body temperature, touch, wind and sunlight. Accordingly, the brand has developed innovative color changing backpack, scarf and phone case. An Italian alligator-skin shoulder bag has been developed with environmentally-responsive ink that turns to black in winters, red in springs, blue in the summers and green fading to red in the autumn thereby eliminating the need to buy different colored handbags to match clothing's colors and seasons.

4.8 Sustainable accessories

4.8.1 Introduction

Accessory designers have also been working towards sustainable alternatives to reduce the footprints in the manufacturing supply chain. Accordingly, they have been diverting attention towards upcycling and recycling of accessories. A curated assortment of vintage and designer handbags is being offered by Rebag thus eliminating wasteful by products in the handbag making.

A footwear brand Veja that also design for Duchess of Sussex are pioneers in practicing fair trade and commit to social and environmental responsibilities. The brand sources rubber from communities in Amazon rainforest.

The brands in the pursuit for effective waste management follow the concept of circular economy and prefer sustainable supplies for their merchandise. The aim of circular fashion is waste reduction by maintaining the consumption and production loop of materials as long as possible. The brands are thus focusing on recycling and resale of clothes no longer in use (Figure 4.6). The designers thus should create their lines from recyclable, sustainable raw materials so that the produced merchandise can offer multiple life cycles. Recycling of apparels can be more convenient if lesser seams are used initially in garment construction and lesser waste is produced as the garment is to cut down into individual pattern pieces.

Brands prefer going in business with suppliers and vendors who adhere to codes of conduct and supply sustainable, eco-friendly raw materials. Accordingly, brands have set up a system of pre-evaluation and rating for suppliers based on their willingness to adopt sustainable measures, technical capabilities, product quality, customer satisfaction and innovative approach.

The consumers' awareness on environmental issues further prompts the brands to adapt to the changing environmental and social dynamics. Brands are marching towards sustainable concept throughout the development process of apparels and accessories to win over consumers trust and enjoy the stature of eco-friendly brand.

4.8. 2 Sustainable accessory brands

A number of successful brands are catering to up market consumers producing value added luxury goods with high environmental value and low production cost by following sustainable concepts.

Insecta, a Brazil based footwear brand uses vintage clothing and recycled plastic (PET) bottles to develop vegan and ecological shoes. Furthermore, the company follows other sustainable strategies, like planting trees to neutralize the CO2 emissions and changing the shipment and delivery modes to bicycle which is by far the most eco-friendly alternative.

Vintage for a cause is a project that is actively involved in transformation of used fashion clothes into vintage apparels and accessories. The revenue generated from the sales of developed vintage clothing is used under the aegis of project to support professional unemployed females.

Filippa K. is another brand pioneering towards effective waste management by selling their used clothes in the brand's retail outlets to encourage shoppers to purchase second-hand clothes. The consumers are further offered with long-term warranties for free repair or replacement of apparels, offering repair, upcycling and providing instructions for repair and upcycling.

Doodlage, an Indian sustainable fashion brand work with eco-friendly fabrics like organic cotton, corn fabric, banana fabric to create innovative apparel designs. They basically work on the concept of recycling, upcycling and zero waste and thus source the fabric from retailers and large manufacturers that deem their unused, discarded, post-cutting fabrics as the waste.

Textiles cast off primarily end-of-season excess and fabric pieces leftover during the cutting room operations in Cambodian factory were utilized for designing ethical women's clothing and accessory line Tonl by designer Rachel Faller.

Apparel and accessory brands have also been experimenting with deadstock. One such apparel brand Reformation uses deadstock as a source of novelty, specialty fabrics, embellished fabrics and faux fur in their apparel lines.

Recode is a Korea based accessory brand that specializes in upcycling. A range of apparels and accessories like jackets, bags and laptop cases are designed by the brand from industrial waste like seat covers, airbags, and fabric linings.

Colorful backpacks, satchels and laptop sleeves are being designed from recycled fabrics by Suave, a Kenyan brand. The materials for their accessory line comprise of off cut fabrics and discarded, waste leather sourced from local second-hand traders, factories and tanneries.

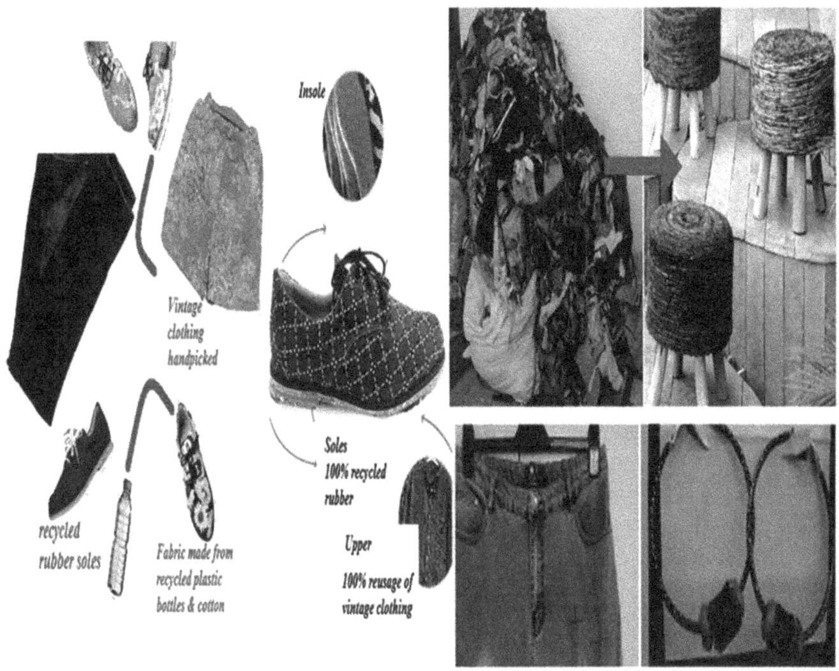

Figure 4.22 Circular economy followed by apparel & accessory brands

5

Jewelry Classification, Designing, Manufacturing & Styles

5.1 Introduction & significance of jewelry

Jewelry is considered to be an embodiment of self-expression and adornment since the time immemorial. Shells, animal teeth, beads were used as adornment by mases in earlier civilization. However, there was yet another dimension attached to jewelry as a means of capital for owner or used as currency. Thus, apart from adornment and economic aspects, jewelry has functional, symbolic, religious membership and status aspects associated with it.

The stature of jewelry is not merely confined to subordinate, aesthetic role to apparels however, kamarband, belt buckle, ornate pin, cufflinks serve functional fasteners along with their aesthetic appeal. Jewelry has become an indispensable component of fashion industry owing to its role in denoting status, utilitarian purpose and surface ornamentation.

5. 2 Jewelry classification & materials

5.2.1 Jewelry classification

Jewelry can be categorized based on the exclusivity and intricacy of design, raw materials and styles. Jewelry is classified into fine, bridge and costume jewelry. Precious metals namely gold and platinum, and semi-precious metals are used for designing fine jewelry. Accordingly, this jewelry category is the most exorbitant jewelry range and caters to elite, high end consumers. Bridge jewelry is available at lower price range compared to fine jewelry owing to usage of less expensive raw materials in the making of former. 10 karat gold, sterling silver, gold plated sterling silver – vermeil are the commonly used materials for bridge jewelry. Apart from 10 karat, 14 and 18 karat gold in smaller quantities, mixed with larger proportion of sterling silver is used. Bridge jewelry made of sterling silver is protected by tarnishing by electroplating rhodium on silver. Crystals, cut glass and semi-precious stones like amethysts, opals, cultured pearls, carnelian, turquoise, coral also find application in bridge jewelry.

The third category of jewelry – fashion or costume jewelry targets the teen and twine population who prefer less expensive, mix and match jewelry to be adorned for casual, daily wear. Accordingly, jewelry makes use of inexpensive metals, wood, plastic, glass, shell. The designs and styles can be imitated from fine, bridge jewelry in less expensive materials thereby making the jewelry available in bulk and low-price points.

The collectible fine jewelry from past decades and centuries include vintage, estate, antique, period and heirloom jewelry. The condition, raw material and rarity of designs determine the exquisiteness and value of vintage jewelry. A range of vintage fine jewelry pieces are exhibited at auction houses, estate sales and even available for purchase on online jewelry selling sites and jewelry collectors for vintage jewelry lovers.

The authentic or reproduction of fine pieces of vintage jewelry are adorned by celebrities and elite segment of society to display their fascination for vintage jewelry and to complement their attire.

5.2.2 Materials for jewelry

The selection of materials for jewelry depends on exclusivity, style, price point of the designed ornaments. The most expensive metals and precious gemstones are chosen for designing fine jewelry while raw materials for bridge and fashion jewelry that primarily focuses middle income consumers utilize semi-precious stones, metals, their alloys and simulated stones in their making. The precious metals utilized for fine jewelry include yellow gold, white gold, platinum family metals like palladium. Likewise, the precious gems such as diamonds, sapphires, rubies, emerald, pearl find application in exclusive, exorbitant range of fine jewelry.

5.3 Jewelry designing

There is an ever increasing demand and the personal pursuit for fashion jewelry. Jewelry designers while designing are inspired from various sources like flora, fauna, geometrical patterns, varied cultures and histories like hippi, bohemian, royal families, mythological characters, art, technologies and scenic locations. The most recent inspiration for designers has been technology as they also seem to be smitten by latest technological innovations in textiles and other interdisciplinary fields ranging from wearable electronics, camouflage, smart textiles to smart phones, smelling and sensing clothing, robots, automation, outer space, 3D printing, nanotechnology and so on. Accordingly, the essence of technology is reflected in the jewelry designed by designers.

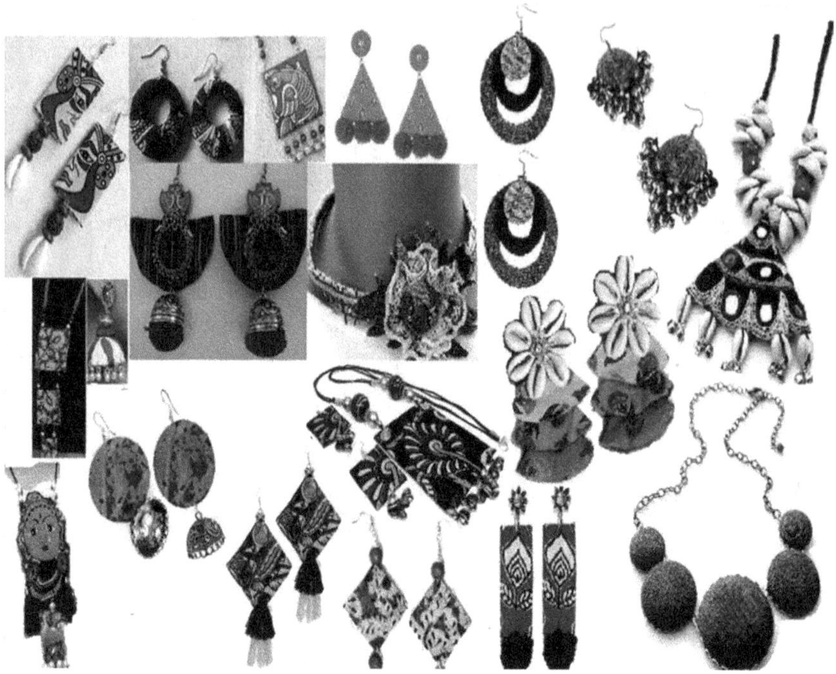

Figure 5.1 Fashion jewelry

A fine piece of jewelry with stones and settings resembling satellites, asteroids, meteoroids with minimal use of silver toned metals like platinum, palladium reflect the designer's inspiration and fascination from space and exploration by space scientists. Designers may create abstracts taking inspiration from nature and various other sources like chinese, medieval, free forming, crumpling, folding patterns from architecture, sculptures, museums, calligraphy and vintage jewelry thereby creation fusion of conservative and contemporary ideas and reinterpretation of ancient artifacts and art forms.

Abstraction involves utilization of original forms in bits and bytes and thus combining or altering the original inspiration source using their own design intuitiveness and creative instincts to create a new design concept. Designers should meticulously consider ready to wear fashion, street fashion trends and sudden, unforeseen trends while designing so that the designs have a glimpse of these trend categories in the upcoming season.

Jewelry designers keeping a watchful eye on ready to wear fashion can create and launch a new product line six months prior to ready to wear collections to make appearance in the retail outlets. Street fashion trends usually last for several seasons and are picked from street fashion. Thus, designers can readily take inspiration and create jewelry merchandise on the same lines. Movies, charity events, celebrity performances inspire consumers to adapt the styles in their lifestyle if they find the trends relatable. Thus, designers can explore sudden, unforeseen trends that hugely impact the public worldwide and can come up with such jewelry designs that can become instant hits.

Designers are also attempting to incorporate sustainable concepts in their design concepts. Accordingly, they are experimenting with various organic alternativesand unconventional fibres like hemp, bamboo and recycled materials for different accessory categories. Jewelry is not far behind with designers exploring the utilization of upcycled and recycled materials in their designs. Haiti-based designers have been designing bracelets made completely with 100% recycled materials like newspapers, plastic water bottles. The initiative has supported the locals of Haiti financially and has created innumerous employment opportunities in jewelry industry.

Designer has to ensure that his hand drawn sketches take into account the forms, colours, print and production possibilities required to create the prototype jewelry.

Apart from hand drawn sketches, digital sketches used by computer intervention can make the designers task simpler and quicker. Adobe illustrator CC, Adobe InDesign CC, Adobe photoshop CC, Sketch up, free design software like GIMP or blinder can be used to make technical drawings of the intended jewelry piece. Digital impression of the fine jewelry piece (which serves as a virtual reality) can be created using the softwares.

5.4 Jewelry manufacturing

The jewelry manufacturing process includes not only use of advanced technologies but also dedicated research.

5.4.1 Traditional techniques

The production of customized jewelry pieces begins with the approval of the designer's sketch by the manufacturers. Transformation of designer's concept and sketch into fine piece of jewelry involves techniques like handcrafting, die striking and casting.

The traditional and the most popular age-old technique of jewelry making is lost wax casting which involves carving of prototype in wax. The wax prototype is encased in a hard substance like plastic or ceramic followed by firing the encased wax in furnace for several hours. The firing of wax causes its vaporization and ultimately in creation of cavity from the burned-out wax. Molten metal is poured into empty space (cavity) and allowed to cool and solidify thus reproducing the hand carved wax model. Single or multiple replicas of pieces can be made using rubber or plastic molds that can be recreated from metal models.

The automated and manual techniques employed which render the manipulation, formation of materials, making metals workable and converting them into desirable shapes for jewelry making are as follows:

Fusing - the technique is used to join two metal pieces together by application of intense heat that liquefies the metals. The technique eliminates the use of any adhering material for joining. The success of fusing depends on precise control of temperature and amount of heat supplied to metals.

Soldering - The process uses a separate metal like silver of lower melting tempertaures. However, color and strength should closely match the two metals that will be joined. Solder itself concealed, serves the purpose of fusing the two pieces of precious metals together or can hold the ornamentation onto the backing. Type of solder used and size of pieces to be joint determines the tools required for holding the two metallic pieces together. Accordingly, wire, steel pins, soldering blocks and plaster can be used for holding the pieces with precision.

Drawing wire jewelry pieces like ear loops, rings, bracelet demand drawing the metals into wires to be transformed into varied styles.

Drawing process involves reduction of heavy gauge wire into predetermined thickness. A device known as drawplate composed of different sized holes is used to reduce the wire to the required thickness. The thick wire is forced through openings of bigger diameter and drawn with pressure so that as it exits the opening of smaller diameter on other side, its thickness is considerably reduced.

Rolling metal is a technique of transforming heavy metal piece into sheet or wires of required thickness via a specially designed equipment called rolling mill.

Metal bending involves the use of square nosed pliers and brakes to bend the metals depending on width of metals. Wider metal piece is bent into sheet by means of brake while specific curve or angle can be achieved on less wider metal pieces by square nosed pliers.

Repousse is preferred over bending when instead of angular bending, metal requires to be transformed into three-dimensional shape. The process is accomplished by hammering a flat metallic piece into hollow, three-dimensional shape as per jewelry design requirements.

Cutting can vary depending on the portion of the metal to be cut and type of cut required. Accordingly, various cutting tools like shears, saws, drill, chisels can be used to cut the metal and render it workable.

Shears can cut through the metal; saws are used for making more precise cuts in the metals while drills are used for drilling holes in the metal. Likewise, chisel is a preferred cutting tool for metal carving.

Hammering imparts different textures, dents on the metallic surface as the metal is beaten with hammerhead.

Antiquing can be credited to heirloom or antique look to jewelry by imparting a darkened finish on the metal.

Engraving process creates intricate designs on the metal's surface by cutting tiny grooves into it.

Chasing is similar to engraving and utilizes hammer, punch to tap designs into metals.

Florentining involves covering the metallic surface by the minute scratches cut into metal as the design feature.

Etching is characterized by recess creation on metal surface by application of acid that cut away some specific area of metal, which appear as recess, dent or a cavity.

Die striking is used for bulk production of jewelry and involves the use of metallic mold or form. The metallic molds are placed between dies where hydraulic press operating at extreme pressure, squeezes metal mold into desired shape. The technique results in durable, high quality settings and generally preferred for rings.

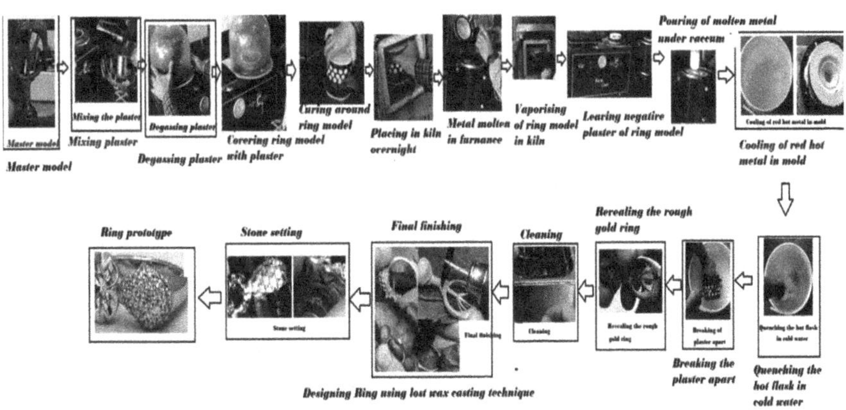

Figure 5.2 Lost wax casting technique for jewelry designing

Casting begins with the formation of plaster, sand, wax mold or the cast into which the molten metal is poured. The mold on immersion in cold water breaks leaving the cooled, solidified metal in the shape of mold. Eventually, cleaning and polishing of the piece is accomplished.

Hand manufacturing technique is preferred for exclusive, customized jewelry pieces intended for elite consumers. A high-level of workmanship and processes like sawing, drawing, hammering, soldering results in the most exclusive and fine jewelry piece.

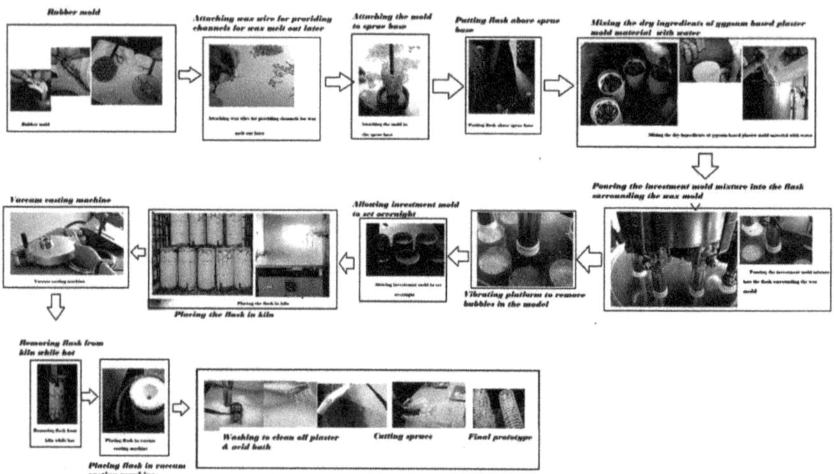

Figure 5.3 Creation of jewelry prototype using rubber molds

5.4.2 Computer aided manufacturing

The traditional jewelry making process has witnessed a major transformation with the advent of CAD, CAM technology in jewelry design and development. The drawing and drafted models of jewelry pieces intended for mass production becomes less time consuming and precise with the inception of computer aided design. The creation of prototype before bulk production of jewelry is crucial for effective utilization of resources and raw materials. Computer Integrated Manufacturing (CIM) technique enables rapid 3D prototype creation thereby making the approval and feedbacks from jewelry brands and retailers quicker. Increased efficiency and precision in model designs is ensured with computer intervention in designing and prototype creation.

Plastic, wax fine jewelry casts can be created with greater level of precision using laser technology as compared to traditional methods of jewelry making.

Refined piece of jewelry is the result of fine workmanship of metalsmith who pours the molten metals into created mold or cast.

Karat quality and manufacturer's registered trademark are stamped on gold jewelry by goldsmith and jewelry manufacturers for updating the consumers about the material content of jewelry that they buy.

5.4.3 3D Printing Technique

3D printing is taking the jewelry industry by storm where a digital file can be converted into a tangible ornament.

3D printer creates a piece of jewelry layer by layer. Nylon powder possessing the ability to melt on heating and then hardening on solidification is utilized for 3D printing. Commercially, the printer with laser technology ensures that laser melt the nylon at exactly precise spots, adding new layers of nylon powder. Thereby, the laser melts the newly formed layer onto the previous one resulting in formation of article layer by layer.

Apart from nylon, gold, titanium and bronze can also be used to create an article layer by layer using this technology. New avenues in jewelry designing and production have emerged with the advent of digital design and 3D printing techniques. The technologies thus enable innovative, latest customization options along with the precision and less time, labor intensive processes compared to the traditional age old techniques of handcrafting and lost wax casting.

Digital workflow empowers the designers to utilize CAD software tools for digital design creation and production of printed patterns using 3D printers which can be used for casting in mold. The burnout of the pattern is followed by process similar to traditional casting technique. The design created by this technique can be easily preserved, modified and recreated as per requirements. Personalized creation and customization no longer is an expensive privilege but is a value added service offered by jewelry brands owing to 3D printing technique used for jewelry designing.

Jewelry retail has benefitted largely by 3D printing because of the capability of technique to create complex geometries and intricate detailing on smaller jewelry pieces like earrings. Furthermore, 3D printing offers flexibility of designing abstract shapes and interlocking structure with considerable reduction in the weight of the jewelry.

SLS, Jet fusion technologies work with plastic powder thereby making the jewelry light weight. Another approach of creating lightweight jewelry is filling the 3D printed jewelry with lattice structure. The lattice can be incorporated in such designs and ornaments which do not require solid interiors like brooches, necklaces, etc. The customization and mass production of jewelry designs can both be benefitted from digital tools.

Wax patterns required for lost wax casting technique can be produced using vulcanized rubber molds. However, the master pattern is still made by traditional hand carved wax pattern. RTV (room temperature vulcanization) molds and high temperature vulcanized rubber molds that are used as master molds can be produced by 3D printers.

The quality of master mold obtained by 3D technique is so smooth that it eliminates any further need of finishing. 3D printing enables multiple patterns to be made simultaneously in a short time span thereby resulting in reduced costs and lead time compared to traditional pattern making techniques like aluminium mold for casting.

Moreover, 3D printing facilitates multiple designs to be produced in single print. Thus, the technique is promising and ensures cost competitive pricing for low production volumes. Investment casting and direct printing are the two methods that 3D printing technique utilized for jewelry design.

Investment casting is the commonly used technique for jewelry making. It involves processes like pattern formation, mold assembly, shell building, burnout, knock off, cut off and finishing.

Pattern formation - 3D printing enables direct printing from wax or castable resin unlike the conventional techniques of pouring casting wax into metal mold.

Mold assembly – involves assembling the molded or printed pattern onto a casting tree enabling multiple components to be casted in one go.

Shell building – involves submerging the entire assembly in slurry several times. Ceramic outer layer is formed over the pattern as the slury coating is left to dry and solidify.

Burnout – this process creates a hollow negative mold (cavity) by placing the structure inside furnace so that wax/resin structure is melted/burnout.

Pouring – The removal of original pattern material from ceramic negative is followed by pouring the casting material into the mold. The casting material is allowed to cool and solidify. *Finishing* stage involves casting the parts in brass and then electroplating with precious metals.

Knock off - involves removal of the outer ceramic mold by knocking off the outer shell by vibration.

Cut off - The individual cast items are separated or cut off the mold tree after the complete removal of ceramic shell.

Finishing - Traditional jewelry finishing technique finish and smoothens the cast parts.

Direct printing utilizing metal powder is a less popular 3D printing technique deployed in jewelry designing.

Gold, silver, platinum alloys are used for printing the jewelry parts. The process requires post processing operations like finishing and turns out to be an expensive option compared to investment casting as the former requires precious powder management.

Despite the innumerous advantages that 3D printing offers in jewelry designing, many consumers and designers are reluctant to adopt the technology as the technology compromises with the hand crafted aspect which has been associated with jewelry designing since ages.

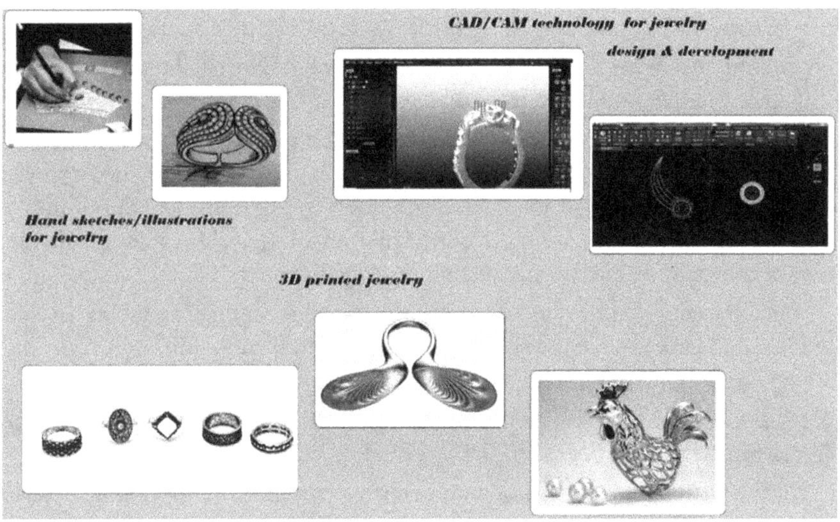

Figure 5.4 Jewelry design

5.5 Stone settings for jewelry

Metals and stones are often combined together to design exclusive jewelry pieces like rings, pins, bracelets, tiara, etc. The exclusivity and value of stone, its durability, security and price point of the finished ornament dictates the stone setting chosen for jewelry designing. The stone settings used can also be chosen based on the usage of the jewelry - i.e., its handling and level of abrasion it will receive, individual preference of consumers for made to measure ornaments and the current trends.

The frequently used stone settings in jewelry designing include tiffany, butter cup, illusion, pave, channel and bezel settings.

Tiffany or the prong setting is frequently preferred for diamond engagement and wedding rings. The setting features prominent display of stones without much intrusion of the metallic component. The setting utilizes four to six prongs or v-shaped projections pressing against the stone and securing it in its exact position.

Butter cup head setting closely resembles the prong setting as this setting too utililizes prongs for affixing the stone firmly. The six prongs in the setting rise from scalloped base and give the illusion of a flower bud. The butter cup stone setting find application in pendants and earrings.

Flush or bezel setting is characterized by a narrow band of precious metal encircling gemstone and holding it in place.

Illusion setting is similar to flush or bezel setting however the metal encircling the stone in this case is decorative, textured thereby rendering illusion of larger stone than its actual size.

Cluster setting is generally preferred to achieve the look of symmetrical pattern formed by numerous tightly arranged stones in the finished product. One larger central stone may be surrounded by upto six smaller stones in cluster setting. A more dramatic and visual impact is achievable by stacking two or more tiers of settings, one on the other.

Pave setting features closely set gemstone concealing the metal between them. The blank area in the setting are tooled into little beadlike surfaces thus surrounding numerous small stones which are paved into this area.

Channel setting is characterised by setting closely spaced stones (concealing metal in between) into grooves or channels.

Tension mount setting is devoid of any prongs. The mounting exposes diamond or other precious gemstone to more light bringing out the colors and brilliance unlike in prong or bezel setting that often conceals part of the stone. The ornaments designed with this stone setting are comfortable to wear owing to absence of prongs which are prone to scratching or catching on clothing. The setting is durable and stronger when compared to other stone settings.

Inlay setting this stone setting involves positioning the stone in recession thus ensuring that stone's surface is not raised. The setting work well with unfaceted stones like lapis lazuli, turquoise, opal, etc.

Paste setting is routinely used for inexpensive fashion jewelry. The setting involves temporary securing of stone by glueing into setting unlike other stone settings which ensure a more secure anchoring of stone to metal. The setting is obviously less durable and there are chances of stone becoming dislodging and falling out as the jewelry collides any hard surface or is subjected to rough handling.

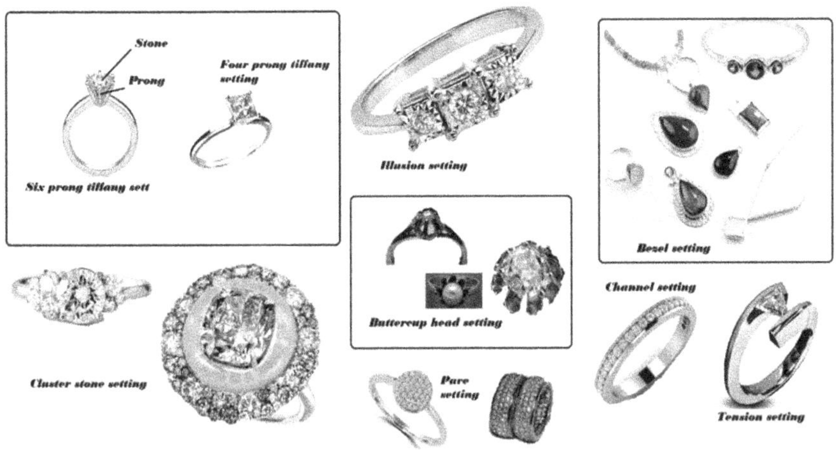

Figure 5.5 Stone settings for jewelry

5.6 Jewelry styles

An array of fashion, costume and bridge jewelry styles are presented by designers to cater to the requirements of consumers. The popularity of a particular style is largely governed by its decorative or functional attributes, materials i.e., metals and stones utilized in manufacturing jewelry, the current trends in apparels and whether the style is a classic or a fad.

5.6.1 Necklace styles

Necklace styles vary according to their length ranging from those that tightly fasten around the neck to the ones that hang down the waistline. Accordingly, necklaces are classified into choker, jewel, chain, matinee and opera based on their length.

Chokers - the length of chokers vary from 14-15 inches and they usually fit snuggly around the neck. They can be single or multistranded.

Bib- A variation of choker is bib also referred to as collarette that features multistrands and extends upto the base of the neck. The short necklace style features a dangling or flowing front ornament.

Dog collar choker - the necklace closely resembles the dog's collar and is a studded style tightly fitting around the neck.

Jewel - is longer than choker and hangs at the throat base on the collar bone.

Chains - the longer length necklace varying from 16-24 inches is referred to as Chains. The variants of Chain necklace as per link styles include box, herringbone, hollow, rope, serpentine.

Matinee ranges in length from 22 – 24inches.

Opera necklace - the length of opera varies from 30-32 inches.

Ropes - are the longest necklace styles which can be as long as 45 inches and generally wrapped around the neck giving impression of wearing multiple necklaces of gradually increasing lengths. *Lariant* - Another variant of rope necklace is lariant necklace. The necklace style features a cord like cylindrical chain that is looped multiple times around the neck.

Furthermore, the necklace styles can also vary in terms of materials used for their construction. Traditional, exorbitant necklaces are usually constructed from precious metals and gemstones namely gold, silver, platinum, diamond, ruby, sapphire, emeralds, pearls, etc. However, the fashion or costume necklaces utilize semi-precious metals, stones and simulated stones like metal alloys, beads, etc.

Graduated pearl necklace - features pearls of varying dimensions with smaller ones on either side of the clasp with the size of pearls increasing gradually with the largest pearl embellishing the necklace's centre front.

Pendant, lavaliere or drop necklace - a necklace is characterised by small ornament such as a locket, precious stone suspending on a delicate chain of either gold, silver or any semi-precious metal.

Slide, zipper or Y necklace - has chain hanging below the pendant thus giving the illusion of English alphabet Y.

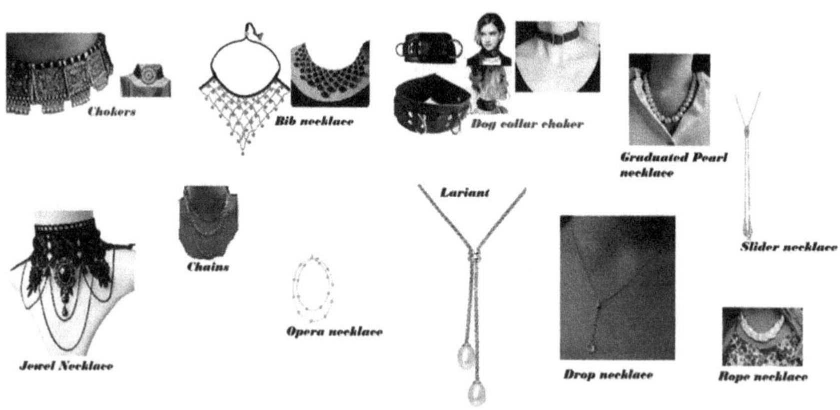

Figure 5.6 Necklace styles

5.6.2 Earing styles

Earrings are designed in a range of materials and styles to match the attire. They can be worn in matched pairs, unmatched pairs, single earrings, multiple earrings on one ear to garner hip-hop, avant-garde look.

Earrings can be categorized into clip-ons, screw back and pierced depending on the manner of affixing them to ears.

Clip on or clip back are best suited for users with unpierced ears. They attach to the ear via spring hinge clamping on the ear lobe.

Screw back earrings like clip on do not require ear piercing. The earrings are held in place by tightly screwing back.

Pierced earrings require piercing in the ear through which post or wire can be inserted. Pierced earrings are equipped with removable back attached to post that ensure firm grip of earrings onto the earlobe. Any chances of infection to wearer's ear are subsided by use of gold, surgical stainless steel posts in the pierced earrings.

Magnetic earring fastening require magnetized backs for holding the earrings front in place. The earrings do not require spring hinges or piercing.

Some other common earring styles are discussed in the following section:

Button earrings as the name suggest closely resemble the buttons owing to their small, round, flat, ball shape. The earrings fit against the earlobe designed in a range of materials like pearls, precious metals, wood, beads, etc. In contrast to buttons, drop earrings extend from the earlobe. They feature a button style post with a dangling bead, jewel or pendant.

Chandelier gives the illusion of miniature ornate lighting features and are designed in wide variety of shapes and materials.

Changeable or earring jacket features an outer disc (a precious metal circlet, diamond stud) surrounding a button style earrings. The latter can be changed as per the wearer's requirements.

Ear cuff can be worn through piercing in the ear or clamped over ear's upper back edge. The ear cuff resembles a miniature wedding band and designed as a spliced, circular metallic band.

Hoop also referred to gypsy is characterised by circular wire of varying diameters inserted through pierced holes. It can also be attached via button style post. Hoops are also designed with dangling charm or bead swinging freely on the wire.

Studs are similar to button earrings and are inserted into pierced earrings by piercing gun. Studs are devoid of any dangling components and feature a precious metal ball or gemstone sitting against the earlobe.

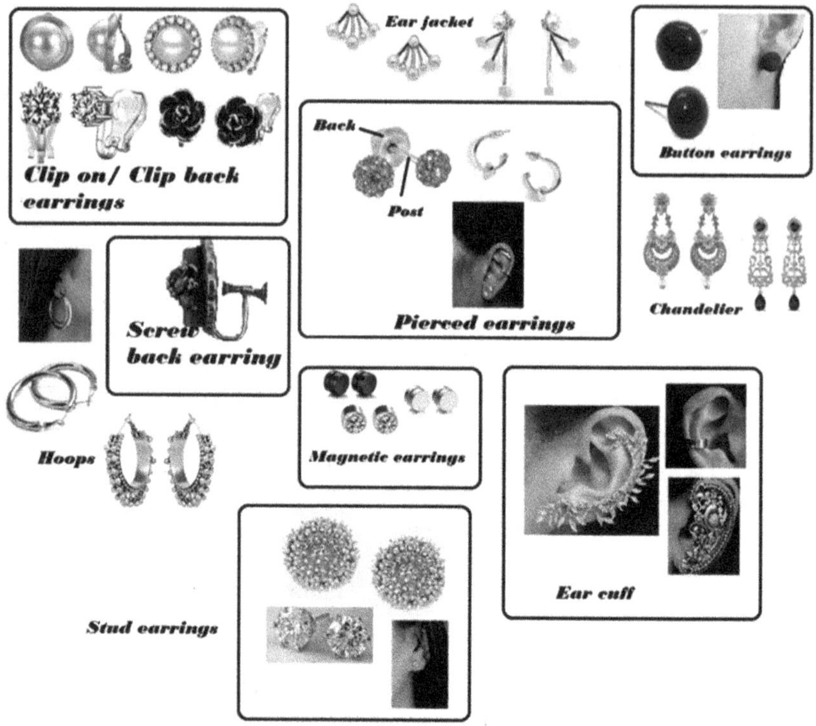

Figure 5.7 Earring styles

5.6.3 Bracelet styles

Bracelets are the decorative bands usually worn around wrist, arm or ankles. Band encircling the upper arm is referred to as armlet or arm band. The bracelet encircling the ankle is termed as ankle bracelet or anklet.

Bangles also referred to as hoops are worn as single entity or in multiples reaching from wrist to forearm. Gold, silver, wood, plastic are the materials of choice for designing narrow, rigid bangles.

Charm is a durable chain link bracelet with a suspended component that can have sentimental, symbolic significance. A variety of images like those of animals, hearts, initials and religious symbols are used as pendants. Gold, silver, precious and semi-precious stones are generally used for designing charms.

Cuff is an elliptical, wide and rigid bracelet with back opening enabling convenient doffing and donning of the bracelet into the wrist. The cuffs can be plain, textured or the wide face of bracelet can be stone embedded or inset with a watch's dial.

Tennis or diamond bracelet is characterised by small, faceted diamonds, sapphire, emerald or any other gemstone set into narrow band or bracelet.

Slave ring also referred to as bracelet ring is combination of ring and bracelet. A chain or medallion connects the ring and bracelet on the back of hand.

Toe ankle bracelet is a toe ring that connects top of the foot to an ankle bracelet.

Contemporary designers prefer a variety of materials namely plastic, brass, wood, colored beads, glass to design exclusive fashion and costume bracelets.

5.6.4 Brooches, pins and clip styles

Brooches, pins and clips constitute the fine jewelry featuring a pin backing for their firm securing on garments or accessories. These go quite well with hats, scarves, handbags or commonly preferred on necklace, shoulder or waistline of women attires.

Apart from pins, hinged clips may also be present for securing them onto apparels. The varied styles of brooches, pins and clips are discussed in the following section:

Scarf clips are used to secure the scarves. Scarf clip is a functional brooch which enables scarf ends to be threaded and holding or securing by means of dual rings at its back.

Brooch or broach is an ornamental pin backed jewelry that is preferred to adorn the look of draperies- like sarees, on necklines or waistlines of an evening dress, or on lapel of men's blazer. Rectangular-shaped brooch is referred to as bar pin.

Lapel pin is a smaller version of brooch that can be worn on the lapel of a blazer or a jacket. Lapel pin features a safety pin backing or a single sharp prong poking through fabric with a removable back.

Hat pin is considered to be functional accessory worn along with hats. Hatpin is characterised by small end piece and an ornamental head on a 9 to 12 inch pin. The pin is made to insert into side crown of hat back and finally emerges towards the crown's back. Hat and hatpin are held in place by fastening end piece on the sharp ends.

Stickpin is a smaller version of hatpin. It is composed of a small ornament fixed on the top of vertical pin. The pin is shorter than hat pin and is generally

3 inches long. The end piece fastens the stick pin as the pin is inserted through lapel fabric.

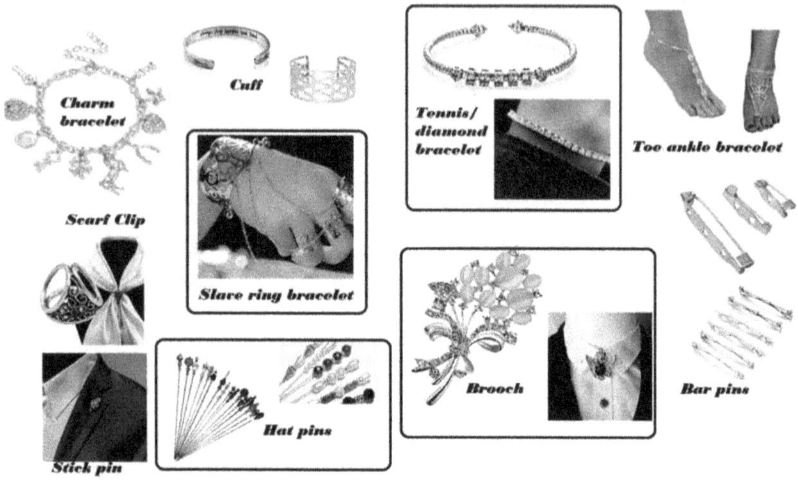

Figure 5.8 Bracelet & pin styles

5.6.5 Ring styles

Rings are designed in a wide range of materials, stone settings and styles depending on consumers' sentiments and preferences while making purchases. Engagement, bridal set and wedding bands are exclusively designed for bride and groom with the precious metals and stones namely gold, silver, platinum, diamond, pearl, etc. The rings are available at high price point owing to the best quality raw material, fine workmanship and customization. Rings are also worn as a metaphor or expression of several emotions, team spirit, belongingness, etc. Accordingly, rings with names of schools, graduation date, a winning teams and an organization's name embedded in gold, silver or any other metal are designed as class, championship and fraternal rings. Likewise, signets are other large sized rings that bear the wearer's initials.

Cocktail or dinner rings comprise of cluster of gemstones piled or swirled resulting in an attractive motif or design.

Nuggets are the rings which have emotional connect as the wearer prefer to use old metal or jewelry (like from mother's vintage collection). The molten gold from old ornaments in its irregular shape like chunk of gold ore can be inset with diamond or other precious stones.

Spinning belongs to category of fashion jewelry. The ring is composed of movable wire circlet that spins as the wearer show any hand activity.

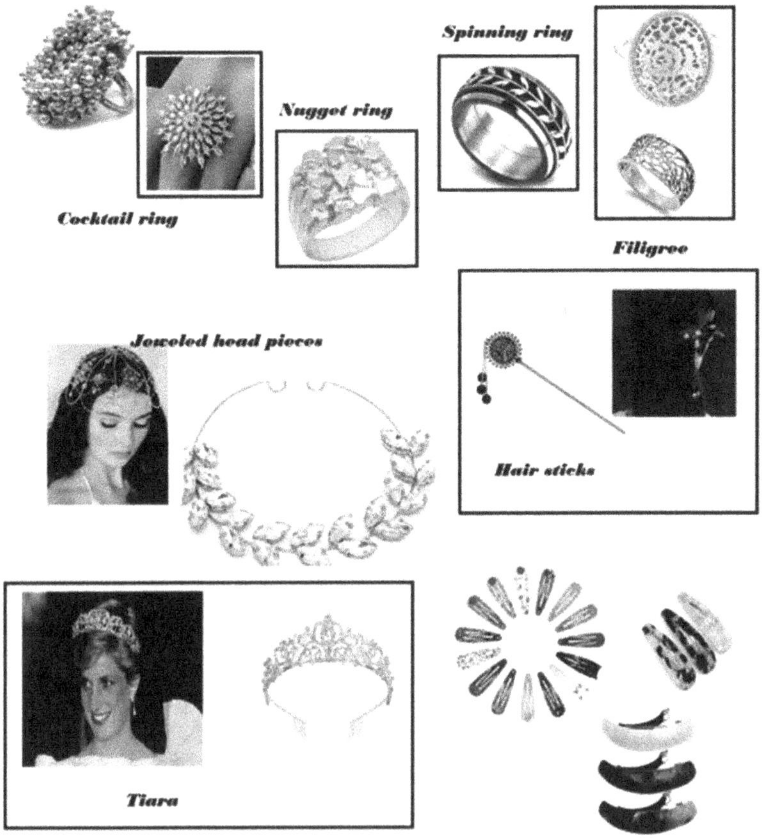

Figure 5.9 Ring styles & hair accessories

Filigree also referred to as Cloisonne is an enamel filled ring. It is thin and twisted wire-shaped ring.

5.6.6 Hair accessories

Jeweled head pieces are hair accessories that include circlet, coronet, crown, diadem and tiara. *Crowns* are larger than circlet and coronets. Tiara generally utilizes gemstones rather than precious metals in its construction.

Hair sticks are used to embellish tresses with ornamental flowers.

Tiara is a royal version of crown that compliments formal gown.

Barettes are clasping hair accessories for firmly securing the hair. Barettes work well individually on a ponytail or in pairs along with braids.

References

http://textilelearner.blogspot.com/2012/02/what-is-textile-fiber-types-of-textile.html

https://en.wikipedia.org/wiki/Fiber

Gini Stephens Frings, Fashion from Concept to Consumer, 9th Edition,2007, ISBN: 978-0131590335

Celia Stall-Meadows, Know Your Fashion Accessories, 2003, Fairchild Books, ISBN: 978-1563672453

Jane Schaffer, Fashion Design Course: Accessories, 2012, BES Publication Co, 978-0764147548

John Peacock, Fashion Accessories: The Complete 20th Century Sourcebook, Thames & Hudson, 2000, 978-0500019979

Jay Diamond, Fashion Apparel, Accessories & Home Furnishings, 2006, Pearson Publishers, ISBN: 978-0131776869

https://sewguide.com/types-of-lace-fabric-trims/

http://www.all-about-leather.co.uk/what-is-leather/leather-types.htm

https://furcommission.com/fur-processing-design/

Sarah Kettley, Designing with Smart Textiles, 2016, Fairchild books, ISBN: 978-1472569158

https://www.truthaboutfur.com/en/fur-dresser

https://furcommission.com/fur-types-in-brief/

https://www.peta.org/issues/animals-used-for-clothing/animals-used-clothing-factsheets/wool-fur-leather-hazardous-environment/

https://www.fibre2fashion.com/industry-article/751/fancy-yarns-for-fashion

https://www.fibre2fashion.com/industry-article/7271/shining-with-metallic-fibre

https://www.power2sme.com/products/details/Yarn

https://bestleather.org/types-of-leather/

https://homequicks.com/different-types-of-leather

https://www.buckskinleather.com/leather-terminology-glossary/

H. V. Sreenivasa Murthy, Introduction to Textile Fibres (Revised Edition), 2015, Woodhead Publishers

https://leatherpanel.org/publications-categories/leather-processing

https://www.loveyourleather.ca/leather-blog/how-leather-is-made-in-23-steps/

https://leatherpanel.org/sites/default/files/publications-attachments/mass_balance.pdf

https://www.hublot.com/en/craftsmanship/design (watches)

https://www.hsamuel.co.uk/webstore/watches/materialsAndComponents.cdo

Index

Chapter 1

Fashion and Accessory terminology	Components of fashion Fashion Cycle, Consumer identification with different phases of Fashion Cycle - Fashion Innovators, Fashion Leaders, Fashion Followers, Fashion Victims & Fashion Laggards Classification of styles – Classics, Fads, Recurring Cycles, Interrupted Cycle Fashion Theories - Trickle-down, Trickle up & Trickle across Theory

Chapter 2

Fashion Accessories Introduction & Classification	Introduction to Fashion Accessories & their importance in apparel industry, Contribution to market profitability Future trends and scope Categories of fashion accessories – Functional & decorative accessories Functional accessories – Footwear, Hats, Handbags, Watches, Eyewear, Scarves, Neck ties, Knitted accessories and Sports accessories. Decorative/Ornamental Accessories – Jewelry, Hair Accessories, High tech accessories. Popular accessory brands & unconventional accessory trends

Chapter 3

Raw materials for fashion accessories	Textile materials – Fibres, yarns & fabrics Fibre properties & classification

	Yarns types & classification based on structure & number of components. Fabrics- knitted, woven and non-woven Non-textile materials – leather, fur, metals & stones Additional components/ sundries

Chapter 4

Construction methods & styles of functional Fashion Accessories	Introduction of functional accessories Required materials and properties of footwear materials Footwear components & construction, CAD intervention for virtual designs and fit customization Men & women footwear styles Hats – Components, materials, construction methods & hat style Handbags – designing, components & materials Construction & styles of handbags Belts – Materials, construction & styles of belts Gloves – Materials, components, construction, styles & glove length Knitted Accessories – Materials, construction methods & styles Smart wearable Accessories & their innovative features Sustainable Accessories & associated brands

Chapter 5

Jewelry Classification, Designing, Manufacturing & Styles	Introduction & Significance of Jewelry Jewelry Classification, Materials & Designing Jewelry Manufacturing- Traditional & Computer Aided Techniques 3D Printing Technique. Stone Settings for Jewelry Jewelry Styles - Necklace & Earing Styles, Bracelet, Brooches, Pins, Clip & ring styles, Hair accessories.

For Product Safety Concerns and Information please contact our EU representative GPSR@taylorandfrancis.com
Taylor & Francis Verlag GmbH, Kaufingerstraße 24, 80331 München, Germany

www.ingramcontent.com/pod-product-compliance
Lightning Source LLC
LaVergne TN
LVHW010635300926
849161LV00005B/945
* 9 7 8 1 0 3 2 6 3 0 0 4 5 *